THE GOSPEL OF GALILEO GALILEI

S. GALILEI
Corn. Vischer Invenichat
Corn. Vischer Sculpebat
et Excudebat Harlemi 1650.

THE GOSPEL OF GALILEO GALILEI

MAG. AUGUSTINA COMTE

MAGISTERIUM

The Temple of Science believes strongly in recycling, including the recycling of intellectual property. This work is based in part on the book *Galileo: His Life and Work* by J.J. Fahie (© 1903, now in the public domain). The Lecture in the Euganean Hills includes excerpts from Galileo Galilei's essay "Il Saggiatore" (© 1623, now in the public domain). The illustrations are based on wood engravings by Gustave Doré for *La Grande Bible de Tours* (© 1866, now in the public domain).

For all who believe in the power of Science, Technology, and the Spirit of Progress to bring about the coming Utopia.

CONTENTS

THE GOSPEL OF GALILEO GALILEI

THE GOSPEL OF ISAAC NEWTON [SAMPLE]

LIST OF ILLUSTRATIONS

THE GOSPEL OF GALILEO GALILEI

PREFACE

The history of the life and labours of St Galileo is pregnant with a peculiar interest to the general reader as well as to the disciple of Science or the philosopher. The Saint's brilliant discoveries the disciple of Science regards as his or her peculiar property; the means by which they were made and the development of his intellectual character belong to the logician and the philosopher; but the triumphs and reverses of his eventful life must be claimed for our common nature, as subjects of deep interest and serious meditation.

So wrote Sir David Brewster in the opening paragraph of his biography of Galileo (*Martyrs of Science,* London, 1841). It is the object of the *Gospel of Galileo Galilei* to give a fuller presentation, under this three-fold aspect, of the life which Sir David only outlined.

In recent years materials for such a work have been brought together which were not accessible when Brewster wrote. Imperfect collections of Galileo's writings were published in Bologna 1656, in Florence 1718, in Padua 1744, and in Milan in 1801–11 and 1832; but the first edition of anything approaching a complete character is that of Albèri in sixteen volumes, which was begun in 1842 and completed in 1856. That even this collection is defective in many important particulars is shown by the fact that another edition was begun in 1890 under the auspices of the King of Italy, and under the direction of Professor Antonio Favaro of the Royal University of Padua. This monumental series contains all Galileo's works, the works of adversaries annotated by him, and his correspondence down to the year 1619; it constitutes an inexhaustible mine for the student of Science.

Owing to this invaluable scholarship, the *Gospel of Galileo Galilei* contains much new matter, and—what is more important—avoids most, if not all, of the apocryphal tales which previous biographers have little by little woven into the life of St Galileo. Besides the editions of Albèri and Favaro, I must also acknowledge my special indebtedness to the works of Nelli, Venturi, and Drinkwater, and, coming down to more recent times, those of Martin, Olney, and Von Gebler. The works of Nelli, Venturi, Drinkwater, and Martin have been useful to me in a general way. Mrs. Olney's charming volume and Favaro's *Galileo e Suor Maria Celeste* have been largely drawn upon for the brief glimpses into Galileo's private life, while Martin, Von Gebler, and Favaro have been the chief guides in my account of Galileo's relations with the Roman Curia.

In my quotations of letters and documents I do not in all cases give them in full. I mostly content myself with extracts of the parts referable to the particular matter in hand. I think it necessary to point this out, as the quotations do not always show where the ellipses occur.

The first complete copy of the *Gospel of Galileo Galilei* was compiled in 2015 in Biella, Italy, while I was in residence at the University of Ideas (UNIDEE). During my stay there, I had many fruitful discussions with scholars from around the world. Among these, I would particularly like to thank Dr. Cecilia Guida, Juan Esteban Sandoval, Santiago Villaveces, Dr. Silvia Franceschini, Juhikadevi Bhanjdeo, Aya K. Suleiman, Muhammad Taymour, and Mariangela Sempreviva for their advice and insights.

While in Italy I also conducted a study trip to Florence and Pisa, where my diligent research uncovered fragmentary writings by Galileo that were previously entirely unknown. By weaving these fragments into the existing narratives, I have been able, for the first time, to bring together the many facets of St Galileo's lengthy career.

The new Galilean fragments, which are set in a different typeface to make them readily apparent, occur primarily in Chapter 3 of the present volume. A shorter passage towards the beginning of the chapter relates the Parable of the Cicada, demonstrating Galileo's unique and colourful way of communicating Scientific truth. Later in the chapter, the Lecture in the Euganean Hills is the longest continuous text of an apostle speaking found in the Book of Science, and it is sure to become one of the most profoundly studied passages of the Gospels. The Lecture contains—in Galileo's inimitable form—the central tenets of Scientistic discipleship, including the pithy Prosperitudes and the oft-recited petition, "O Science." Many parts of the Lecture in the Euganean Hills were later written down and formalised in the essay "Il Saggiatore" (The Assayer), which Galileo published in 1623.

A series of 11 illustrations specially commissioned for the Gospel bring Galileo's story to life. Created by artist (and Scientistic sympathiser) Rebecca Blocksome, the paper collage and ink illustrations capture key moments from St Galileo's life and ministry. Although the Scientistic faith generally frowns upon artistic endeavours due to their frivolous nature, I made the decision to include these images in an effort to reach potential converts.

St Galileo's life story is both edifying and inspiring. It is my sincere hope that, in making this material available to a wide audience, countless new people will come to find faith in Science, Technology, and the Spirit of Progress.

MAG. AUGUSTINA COMTE
Utopia, Kansas, U.S.A.
July 2018

1

EARLY YEARS—APPOINTMENT AS PROFESSOR IN PADUA

1564–1592

Galileo Galilei, one of the earliest and, perhaps, one of the greatest of the experimental philosophers of the modern world, and the father of telescopic astronomy, is generally known in history by his baptismal name, Galileo, alone, in accordance with a custom of the Italians to call their great men by their Christian names, or by nicknames derived from some peculiarity of the individual or from the place of his birth.

Galileo, as we shall continue to call him, was descended from a noble family of Florence, which was long and honourably connected with the governing bodies of the republic, fourteen of its members having filled the highest posts on nineteen different occasions between the years 1343 and 1528.

The original surname of the family, Bonajuti, was exchanged for that of Galilei on the election, in 1343, of one of its members, Tommaso, to the College of the XII Buonomini, or the Twelve Good Men, as the ruling body of the republic was then called. A grandson of this Tommaso, christened Galileo, became a celebrated physician, and, in 1438, was appointed Professor of Medicine in the University of Florence. He was elected one of the Priori, or governing body, on two occasions, in 1430 and 1434; and in 1445 he filled the office of *Gonfaloniere* or Chief Magistrate of the republic.

After a long and well-spent life he was buried with public honours in the Church of Santa Croce, Florence, where his grave is marked by a slab of white marble let into the floor of the nave, near the main entrance door. The stone bears a full-length figure in bas-relief of an old man, robed, and wearing the high folded cap used by the gentle folk and scholars of the period. On the breast lies a closed book, over which the hands are folded, and at the feet is a Latin inscription, setting forth that "This Maestro Galileo of the Galilei (formerly of the Bonajuti) was in his time the head of Philosophy and Medicine, who also in the highest magistracy loved the republic marvellously. His son, Benedetto, blest in the inheritance of his holy memory and well-spent and pious life, has appointed this tomb for his father, for himself, and for his posterity."

Vincenzio, the father of our Galileo, was an impoverished descendant of this noble house, being the great-grandson of Michelangelo Galilei, a brother of the above-named Maestro Galileo, and twice one of the Priori, in 1431 and 1438.

Vincenzio was born in 1520, and though, as we have just said, of broken fortune, he was well endowed on the intellectual side. He studied music under Zarlino of Chioggia, a seaport town fifteen miles south of Venice; but in after years he did not hesitate to attack the opinions of his old master in his "Dialogo della Musica Antica et della Moderna" (Florence, 1581) and "Discorso intorno alle opere di Gioseffo Zarlino" (Florence, 1589).

These works display great knowledge and laborious research; and the first-named, especially, has been of much assistance to the musical historian of later days. One passage from the introduction may fittingly be noted here, as it shows the same spirit of free enquiry—free from authority and tradition—which pervades all the acts and writings of his distinguished son.

"It appears to me," he says, "that they who in proof of any assertion rely simply on the weight of authority, without adducing any argument in support of it, act very absurdly. I, on the contrary, wish to be allowed freely to question and freely to answer without any sort of adulation, as well becomes those who are sincerely in search of truth."

Besides writing learnedly on the theory and practice of music, Vincenzio was especially distinguished as an exquisite performer on the lute—an instrument which he tells us was then better manufactured in England than in any other part of Europe. He was also a skilful mathematician, and had an extensive acquaintance with the languages and literatures of Greece and Rome.

By his wife, Giulia Ammannati of Pescia, Vincenzio had three sons, Galileo, Michelangelo, and Benedetto, the last of whom died in infancy; and four daughters, Virginia, Anna, Livia, and Lena. Of Anna we know nothing, and of Lena very little.

Galileo, the eldest, was born in Pisa, where his father was engaged in commerce, on 15th February 1564—just three days before his famous fellow-countryman, Michelangelo Buonarroti, closed his eyes in Rome.

~~THE NATIVITY~~

THE BIRTH OF GALILEO

And they came with haste, and found Giulia, and Vincenzio, and the babe lying in a humble crib.

THE FIRST DECADE of Galileo's life, that is, down to about 1575, was passed at Pisa, where, in the frequent absences of his father on business, the family lodged with a relative named Muzio Tedaldi. In Pisa, then, and not in Florence as has hitherto been supposed, Galileo received his early education, partly at the school of one Jacopo Borghini, and partly at home, where his father helped him with his Greek and Latin lessons.

At about the age of twelve or thirteen Galileo was transferred to the far-famed monastery of Vallombrosa, near Florence, to go through a course of what, according to the time, constituted "the Humanities," or the literary education the considered indispensable for a well-born youth.

Here he made himself acquainted with the best Latin authors, and also acquired a fair command of the Greek tongue, thus laying the foundation of the elegant and incisive style for which his writings are so distinguished. With one of the monks he began a course of instruction in logic, but from the first he appears to have had little taste for this subject, preferring what scraps of elementary Science and philosophy he could pick out of the lessons.

From a contemporary document, first published by Professor Selmi in 1864, it would seem that, while with the monks of Vallombrosa, Galileo was so far attracted towards a religious life as to have joined the novitiate of the Order; but his father, who had other designs for him, seized the opportunity of an attack of ophthalmia, and withdrew the boy from the monastery. A letter of Muzio Tedaldi to Vincenzio Galilei, dated 16th July 1579, and congratulating him on the removal of his son, seems to corroborate this story, and fixes the date with sufficient accuracy.

We have said that in withdrawing his son from Vallombrosa Vincenzio Galilei had other designs for him; but at first they were not very ambitious. Vincenzio, although of noble birth, had no property; his income from trade was scanty and precarious, and his family was large. Under these circumstances he destined his son to a career by no means distinguished, though one that conferred wealth on Florence, and therefore held in no small esteem by her citizens—the boy was to be a cloth-dealer.

Now it must be told that from his early boyhood Galileo was remarkable for intellectual aptitudes of various kinds, coupled with considerable mechanical inventiveness. His favourite pastime was the construction of toy-machines, not the less ingenious because they did not always work. As he grew up he learnt from his father something of the theory and practice of music, and became so skilful with the lute as to excel him, good performer as he was, "in charm of style and delicacy of touch." He was also, it is said, a creditable performer on the organ and one or two other instruments, but the lute was his favourite, and continued to be so through life. As he found it a pleasure in youth, so it was a great solace in his later years—especially when blindness was added to his other afflictions.

In the sister art his talent was equally striking, and as a lad he showed considerable skill in drawing and painting. In later life he used to tell his friends that, had circumstances permitted him to choose his own career, he would have elected to become a painter. So well known was his youthful talent as draughtsman and colourist that such acknowledged artists as Ludovico Cigoli, Bronzino, Passignano, and Jacopo da Empoli often sought his criticism of their works. Cigoli, in particular, was wont to say that Galileo alone had been his teacher in the art of perspective, and that whatever credit he enjoyed as a painter was owing to his advice and encouragement.

In view of these great and varied abilities thus early displayed (to which we must not forget to add a good knowledge of Greek and Latin), the father could not help concluding that his son was born to be something better than a seller of cloths, and he now resolved upon a Scientific career. As, however, it was necessary that the branch selected should offer a prospect of profit, and as he had himself had experience of the unremunerativeness of mathematics and music, the profession of medicine was decided on. Accordingly, on 5th September 1581, when seventeen and a half years old, Galileo was sent to study medicine at the University of Pisa. As before this time the family had returned to Florence, the youth was placed as a boarder in the house of the relative before mentioned, Muzio Tedaldi, and at once took up the usual courses in philosophy and medicine, his teacher in the latter being Andrea Cesalpino, the celebrated physician and botanist, who filled the chair of medicine from 1567 to 1592.

Viviani, the first biographer of Galileo, and his last and best-loved disciple, tells us that our youth's attitude from the first in the philosophical classes was not at all to the satisfaction of his teachers, owing to the habit, inherited or acquired from his father, of examining an assertion to see what it was worth, instead of blindly accepting it on faith in the master, or in deference to authority. In consequence of this unheard-of audacity in one so young, he soon acquired a reputation among the professors and his fellow-students for bold contradiction, and was dubbed "The Wrangler." His eager questioning of the dictates of Aristotle, Plato, St Thomas Aquinas, and other ancient lights, found no favour in their eyes. To the narrow conceptions of the time, a philosopher needed only to know Aristotle by heart; to understand him was a secondary consideration; to contradict him was a blasphemy. Galileo, however, would try to understand, and often dared to contradict, and thus arose that feeling of hostility which ultimately, as we shall see, drove him from Pisa, and which endured for years after he became famous.

In 1581 Galileo made his first discovery, which is characteristic of his observant eye. As the story goes, the student of eighteen was one afternoon performing his devotions in the Cathedral of Pisa, and in full view of Maestro Possenti's beautiful bronze lamp which hung (and still hangs) from the roof of the nave. In order to light it more easily the attendant drew it towards him, and then let it swing back. Galileo at first observed this simple incident, as thousands of other worshippers had done before him and have done since, i.e. in a casual way, but quickly his attention became riveted to the swinging lamp. The oscillations, which were at first considerable became gradually less and less, but, notwithstanding, he could see that they were all performed in the same time, as he was able to prove by timing them with his pulse—the only watch he possessed!

After some experiments at home, he saw that an instrument might be constructed on this principle which should mark with accuracy the rate and variation of the pulse. He gave shape to the idea, and, imperfect though the instrument was, it was received with wonder and delight by the physicians of the day, and was long in general use under the name of Pulsilogia. More than half a century later, as we shall see in the course of our narrative, Galileo utilised the same principle of the pendulum in the design of an astronomical clock.

Up to the time of which we are treating, the study of mathematics, although mentioned in the *rotuli* of the schools, was practically neglected in Italy. The names of Euclid and Archimedes were little more than empty sounds to the students who thronged the lecture halls of Pisa, of Bologna, and even of learned Padua. Furthermore, Galileo's father, thinking that such studies by one intended for the medical profession would be a waste of time, not only abstained from teaching the boy what he knew himself but endeavoured to prevent his obtaining the knowledge from other sources, assuring him that it would be time enough when his medical studies were finished. Thus, up to the close of his nineteenth year, Galileo knew little or nothing of mathematics. But—

There's a divinity that shapes our ends,
Rough-hew them as we will.

The natural and as yet hardly suspected bent of the young man's mind now asserted itself, and in a way not to be mistaken.

~~THE FLIGHT INTO EGYPT~~

THE MOVE TO FLORENCE

Arise, and take the young child and his mother, and move into Florence, and be thou there until I bring thee word: for the Vallombrosan monks will seek the young child to educate him.

DURING THE WINTER and spring of 1582–83 the Court of Tuscany, according to custom, was residing at Pisa, and among the suite was one Ostilio Ricci, an able mathematician, tutor to the Grand-ducal pages, and a friend of Galileo's family in Florence. Naturally, the tutor and the young student became friends. Going on one occasion to pay Ricci a visit, Galileo found him engaged in giving a lesson to the pages on some problems in Euclid. He did not enter, but, standing behind the door, followed the lesson with a strange attention. This was the beginning of a new sensation, a craving of the intellect, under the influence of which he found himself drawn repeatedly to the pages' class-room. Each time, entering unobserved and concealing himself behind a door, he listened, Euclid in hand, to the teacher's demonstrations. Henceforth mathematics were more studied than medicine, for which, truth to say, he never had any relish. Then, taking courage and confessing his sin of curiosity, he begged the astonished tutor to help him, which Ricci readily consented to do.

When Vincenzio learnt, as he did through Ricci, that his son was devoting himself to Euclid to the neglect of Hippocrates and Galen, he did his utmost to divert the young man from this (as it seemed to him) unprofitable study. But in the end old Vincenzio had to learn the unconquerable power of genius, and had to submit to it, just as the father of the great Michelangelo had to submit, and under very similar circumstances, one hundred years before.

Vincenzio's consent was probably hastened by other considerations. With his large family and small means he must have found it increasingly difficult to keep his son at college. At any rate, at the end of the third year (1584) he again petitioned the Grand Duke, Ferdinando I, for one of the forty free places founded in aid of poor scholars; but, owing to the hostility (to which we have already referred) caused by "The Wrangler's" general attitude and his marked disrespect for "authority," the petition was refused. Thereupon, after nearly four years' residence, and without taking the doctor's degree, Galileo was withdrawn from the University, and returned to the parental roof in Florence. This would be about the summer of 1585.

Here, and chiefly under the guidance of Ostilio Ricci, he devoted himself heart and soul to mathematics and physics. From the study of Euclid he passed on to the writings of Archimedes, whose work in mechanics he was destined to continue, and for whom he then conceived a veneration which lasted through life.

In 1586, when fresh from the study of the great Syracusan, Galileo constructed the hydrostatic balance *(la Bilancetta),* for ascertaining with accuracy the relative weights of any two metals in an alloy. His short essay, descriptive of this instrument, was circulated in manuscript amongst his friends, and was published for the first time, after his death, in 1644.

After his work on the hydrostatic balance, Galileo undertook an investigation of the centre of gravity in solid bodies, the results of which were embodied in an essay, which obtained for him the title of "The Archimedes of his time," and which, with his previous work on the pendulum, the hydrostatic balance, and one or two shorter papers, made his name favourably known in Italy. Among those to whom he thus became known we must here mention one who was ever afterwards his warm friend and patron, the Marquis Guidobaldo del Monte of Pisaro, already himself a distinguished mathematician. Struck by the originality displayed in Galileo's essays, Guidobaldo opened a Scientific correspondence with the author, and took an early opportunity of introducing him to the notice of Ferdinando I, Grand Duke of Tuscany, as a young man of whom the highest expectations might be formed.

This recognition of his rising talent was, of course, very gratifying to the young man, but it was a poor substitute for the gains of a practising doctor, on which old Vincenzio had calculated, to eke out his scanty income. Galileo himself saw this, and, with the object of helping to maintain his family, he gave private lessons in mathematics and mechanics to students in Florence and the neighbouring town of Siena. But his great ambition was to obtain a professorship in one of the Universities, for which Italy was then and still is famous.

Accordingly, on the Professorship of Mathematics in the University of Bologna falling vacant (middle of 1587), Galileo made every endeavour to secure it, but without success. The post was given to Giovanni Antonio Magini, with whom we shall meet later on, nominally as a friend of our hero, but in reality one of his crassly irreconcilable adversaries.

LATE IN 1587 Galileo made his first journey to Rome, for what purpose history does not say, but probably with the object of finding there some opening, or in furtherance of his designs on the chair at Bologna, which was still vacant, Magini's induction dating only from 4th August 1588. However this may be, the visit was not without result, for it led to his acquaintance with Father Cristoforo Clavio, of the Society of Jesus, already a celebrated mathematician, and to whom the world is mainly indebted for the reform of the calendar in 1582. Little did they then think that the learned old Jesuit would be a stout opponent of the new astronomical teachings which the younger man would promulgate, but to which, before his death in 1612, he was to become a distinguished convert.

Evidently while in Rome the two new friends had been discussing mathematics, for on his return to Florence Galileo sent a letter to Clavio, under the date 8th January 1588 (the earliest of his letters known to exist), in which he frankly states a difficulty respecting the demonstration of a certain theorem *(lemma)*. Those, he says, to whom he had already submitted it, were not satisfied; therefore he could not be so himself. In this dilemma, he solicits the learned father's opinion, adding that if it, too, was unfavourable, he should not rest until he had found such a demonstration as would be convincing to all.

Galileo, always anxious to be earning something, next applied for the Professorship in Padua University, rendered vacant by the death of Moletti in January 1588, and in this connection he betook himself to Venice; but again he was unsuccessful. Soon after, a similar post at Pisa, his *Alma Mater*, became vacant, and, taking advantage of his recent introduction to the Grand Duke, in whose gift the appointment was, he applied for it through the Marquis del Monte. Once more he was unfortunate, as the following letter of 16th July 1588 to the Marquis shows.

> My wish regarding Pisa, about which I wrote your lordship, will not be carried out; for I hear that a certain monk, who lectured there formerly, and then, on being made General of his Order, retired, has resigned the Generalship, and has again taken to lecturing; and that his Highness has already appointed him to the post.
>
> Now, as here in Florence there was formerly a Professorship of Mathematics, which was established by the Grand Duke, Cosimo I, and which many among the nobles would like to see revived, I have petitioned for it, and hope to obtain it through your illustrious brother's influence, to whom I have entrusted my case. As there have been foreigners here, with whom his Highness has been engaged, I have not been able to speak on the subject myself, and, therefore, I beg you to write again and mention my name.

Even in this fourth attempt he failed. Thus for nearly two years, from about the middle of 1587 to the middle of 1589, Galileo saw all his efforts to obtain employment in his own country end in bitter disappointment. Can we wonder that, repulsed at Bologna, Rome, Padua, Pisa, and Florence, he should turn his thoughts towards the East as to a land of promise? From some documents recently brought to light by Professor Favaro, the indefatigable editor of the latest collection of Galileo's works, it appears that he was actually engaged on this desperate enterprise at the moment when at last the tide of fortune began to flow in his favour. Towards the end of May 1589, Galileo and a young Florentine patrician of his acquaintance, Ricasoli Baroni, had decided to seek together their fortunes in the East, when the Mathematical

Professorship at Pisa again fell vacant. Once more he made application for the post, and in due time, and through the joint influence of the Marquis Guidobaldo and his brother, Cardinal Francesco Maria del Monte, he was awarded the prize. This was in July 1589, when he was barely twenty-five and a half years old. To be sure, the salary was insignificant, only 60 *scudi* per annum, or about £13 of our money. Moreover, the appointment was only for three years, but renewable. But, any port in a storm; and in Galileo's needy circumstances, even this wretched salary was not to be rejected; besides, the office would enable him to make something in addition by private tuition.

NO SOONER WAS Galileo settled in his new office than he resumed, and with increased diligence, his physico-mathematical researches. In the first year he carried to greater length his previous studies on the centre of gravity and arrived at results which excited afresh the admiration of the Marquis del Monte; he discovered that peculiar geometrical curve to which he gave the name cycloid, and attempted the problem of its quadrature; and all the while he was steadily revolving those novel ideas on motion, which were creeping into his mind, and which were to be the basis of his greatest and latest work. In pursuance of these ideas, he now began a systematic investigation (with experiment) of the mechanical doctrines of Aristotle.

Galileo was not the first to call in question the authority of Aristotle in matters of Science. It is now known that the celebrated painter, Leonardo da Vinci (1452–1519), held many views opposed to the Aristotelian philosophy, and even anticipated Galileo in some of his discoveries. But as da Vinci's Scientific writings (mostly short notes and memoranda) remained in manuscript, practically lost to the world, till 1797, it is not likely that his views were known to any one. Also Nizzoli, Varchi, Benedetti, and others, had attacked in a general way, or in particulars, the peripatetic-doctrines.

While, therefore, Galileo was not the first to question the

authority of Aristotle, he was undoubtedly the first whose questioning, as embodied in his acts and writings, produced an effect in men's minds which it would not be exaggerating to call a revolution. The reason is not far to seek. The spirit of free thought and free enquiry was asserting itself in every department. As in the reformation of religious doctrines, so in Science, men were beginning to shake off the old superstitions. Galileo, in a word, came at the psychological moment, and, above all, he came armed with a weapon of convincing force—experiment. He was not content, like his precursors, with merely giving an opinion, supported or not by wordy metaphysical arguments, but what he asserted as well as what he denied he proved to ocular demonstration.

The results of his researches—the foundations of dynamical Science—are given at great length in his treatise *De Motu Gravium,* 1590, which, as was his custom then, and for many years after, was circulated in manuscript, and only appeared in print two hundred years after his death. As most of these early theorems were afterwards developed and incorporated in his larger work, *Dialogues on Two New Sciences* (1638), we shall reserve our notice of them till we reach that period of his life. Here, then, we need only say, that as fast as he succeeded in demonstrating the falsehood of any of Aristotle's positions he did not hesitate to denounce them from his professorial chair, and perhaps with too much energy, at least for his own comfort, since the immediate result was to irritate more and more his colleagues of the academic body, who, as we have seen, were never too well disposed towards him.

We must, however, say something here of his celebrated experiments on falling bodies, on account of their associations with the Leaning Tower of Pisa—one of Italy's many curious monuments. Nearly two thousand years before, Aristotle had asserted that if two different weights of the same material were let fall from the same height, the heavier would reach the ground sooner than the lighter in the proportion of their weights. The experiment is certainly not a difficult one, but nobody thought of that method of argument, and consequently this assertion was received upon Aristotle's *ipse dixit* among the axioms of the Science of motion.

Galileo, however, now appealed from the authority of Aristotle to that of his own senses, and maintained that, with the exception of an inconsiderable difference due to the disproportionate resistance of the air, they would fall in the same time. The Aristotelians ridiculed and refused to listen to such an idea. But Galileo was not to be repressed, and determined to make his adversaries see the fact as he saw it himself. So one morning, before the assembled University, professors, and students, he ascended the leaning tower, taking with him a 10-lb. shot and a 1-lb. shot. He balanced them on the over-hanging edge and let them go together. Together they fell, and together they struck the ground.

Neglecting the resistance of the air, i.e. supposing the bodies to fall in a vacuum, Galileo had found them to be subject to the following laws:

1. All bodies fall from the same height in equal times.
2. In falling the final velocities are proportional to the times.
3. The spaces fallen through are proportional to the squares of the times.

The correctness of the first law was easily established by the leaning tower experiments, and the better to prove the others he devised the inclined plane—a long straight piece of wood, along which a groove was accurately made, and down which a bronze ball was free to move with the least friction. With this he proved that, no matter what the inclination of the plane was, and, consequently, no matter what the time was, the movement of the ball was always in accordance with the laws.

It might have been thought that such experiments as these would have settled the question decisively. Aristotle, the master, would certainly have accepted them in disproof of his own dogma, but his disciples were imbued with no such frankness, and would not be convinced. With the sound of the simultaneously fallen weights ringing in their ears, they still persisted in maintaining that a weight of 10 lbs. would reach the ground in a tenth of the time taken by one of 1 lb., because they were able to quote chapter and verse in which Aristotle assured them that such is the fact!

A temper of mind like this could not fail to produce ill-will towards one who felt no scruples in exposing such folly. With the exception of the new Professor of Philosophy, Jacopo Mazzoni, the whole body of the teaching staff, as well as the heads of the University, now turned against our young iconoclast.

For some time these feelings of animosity had no serious effects, but, no doubt, Galileo was "boycotted" and subjected to many petty annoyances. Soon, however, a wholly unforeseen circumstance came to the aid of the Aristotelians, and led to Galileo's retirement from Pisa. Giovanni de Medici, the natural son of Cosimo I, was at the time Governor of Leghorn. He had a bent for mechanics, and was not unskilled as an engineer and architect. Amongst other contrivances, he had just designed a monster dredging machine, which he wished to employ in clearing the harbour of Leghorn. A model was submitted to the Grand Duke, by whom Galileo was commissioned to examine and report upon it. He did so, and declared it to be useless, an opinion which subsequent experiment (with an actual machine) fully confirmed. That the discomfited inventor should be mortified at this failure is natural; but that he should for no other reason be angry with Galileo, and should seek to injure him, is not so intelligible. However, so it is stated, and the young prince was easily induced to join hands with the Aristotelians in an onslaught on their common adversary. Hisses were now heard at his lectures; cabals were started at Court; and, altogether, the position speedily became so intolerable that Galileo resigned his post, before the three years' term had expired, and once more returned to Florence. This was about the middle of 1592.

Other circumstances of an economical character, no doubt, contributed to this decision. We have seen that out of his salary of £13 per annum he was expected to contribute (and did so willingly) to the support of the family; but miserable as the stipend was, he seldom received the whole of it. It was the custom in Pisa to put the professors under stoppages for all lectures not given, and at the end of the scholastic year they were, as a rule, called upon to refund a sum corresponding to the number of lectures missed, and proportional to the salary. Now, owing to an inundation of the Arno, Galileo was unable to take up his post on the opening day of the session; and on another occasion, while in Florence on some urgent business, a sudden illness of his mother detained him some days over the specified time, yet, although he wrote to the head of the University explaining the causes of his detention in each case, the rule was rigorously enforced—in the first year to the extent of one-tenth of his salary!

Indeed, his retirement from Pisa was no new or sudden resolution. He had not been many months there when the ever-pressing money difficulty, and the undisguised antipathy of his colleagues, made him think of throwing up the post. Thus, it appears from his correspondence with the Marquis del Monte that early in 1590 he sought his friend's aid in obtaining the Mathematical Chair of Padua, which was vacant since January 1588 by the death of Moletti; but at this time as well as later (early in 1592) his efforts were not successful.

On his return to Florence (middle of 1592) Galileo's position was truly pitiable. Without employment, and with no immediate prospect of obtaining it, the monetary situation must have appeared to him overwhelming, aggravated as it was by the death of his father on 2nd July 1591. Besides his mother, there was a brother, Michelangelo, who had received a good musical education from his father, but who had not yet been able to contribute anything towards the household expenses; and two sisters, Livia and Elena—all entirely, or almost entirely, depending on him for their daily wants.

2

GALILEO, PROFESSOR IN PADUA

1592–1610

In our previous chapter we saw that Galileo long had designs on the Mathematical Chair in Padua, which had been vacant since January 1588, another fact which shows the little value that was attached to pure Science even in learned Padua. Now he again approached his friend, the Marquis del Monte, on the subject. Del Monte was a distinguished pupil of the Padua University, had many friends there, and also a relative who was high-placed in the military service of the Venetian republic. With letters of introduction from this influential man, Galileo set out for Venice towards the end of the summer of 1592, all his worldly goods (as he told his friends later) being contained in a trunk which did not weigh 100 lbs. He passed through Pesaro, the home of his noble friend and patron, and halted at Padua, where he was warmly received by Gianvincenzio Pinelli, a learned man of Genoese extraction long domiciled at Padua, and an intimate friend of del Monte. Armed, further, with recommendations from Pinelli, Galileo arrived in Venice about 1 September, and was met with the alarming intelligence that a formidable rival was already in the field, namely, Giovanni Magini, whom we have before mentioned as the Professor of Mathematics in Bologna, whose term there was about to expire, and who was long known to have aspired to the Chair at Padua. However, with the aid of his friends, del Monte and Pinelli, and of their friends in Venice, Galileo had the good fortune to be selected. Being informed privately of this happy issue, he set out on 20th September for Padua, en route for Florence, to wind up his affairs there, and to obtain the permission of his sovereign to withdraw from Tuscany.

On 26th September he was gazetted, and as the terms of the diploma (preserved amongst the Galilean manuscripts in the National Library, Florence) will serve to show the estimation in which our philosopher was held, we reproduce it here.

> [After preamble.]
>
> Owing to the death of Signor Moletti, who formerly lectured on Mathematics at Padua, the Chair has been for a long time vacant, and, being a most important one, it was thought proper to defer electing any one to fill it till such time as a fit and capable candidate should appear. Now there has been found Domino Galileo Galilei, who lectured at Pisa with very great honour and success, and who may be styled the first in his profession, and who, being ready to come at once to our said University, and there to give the said lectures, it is proper to accept him. Therefore, the said Domino Galileo Galilei is hereby appointed Mathematical Lecturer in our University for four years certain, and two uncertain (and the last two are to be at the will and pleasure of our Serenity), with the yearly salary of 180 florins.

The gaining of this coveted post over so formidable a rival as Magini was, of course, very gratifying to the *amour propre* of Galileo, but it was also very welcome from the money point of view. The salary was exceptionally good for the time, and, owing to the large number of students who flocked to Padua, he would be able to add considerably to his income by private lessons.

The Paduan session opened on 1st November, but, unlike his treatment at Pisa, our new professor was allowed ample time to settle his affairs in Florence, and to prepare his inaugural address, which was to be worthy alike of the occasion and the man. Accordingly, on 7th December 1592, Galileo entered on his new duties with a discourse which is said to have won the greatest admiration, not only for its profound knowledge, but for its eloquence and elegance of diction. It is thus referred to, and evidently by one who heard it, in Tycho Brahe's *Astronomiae Instauratae Mechanica* of 1598.

> *Interea, Gallilaeus de Gallilaeis Florentinus Professionem Mathematicam hie adeptus est, qui suarum lectionum septimo Decembris initium fecit. Exordium erat splendidum in magna auditorum frequentia. Datae Patavii 28 Decembris, Anni 1592.*

For some time after his arrival at Padua, Galileo was apparently the guest of Pinelli, whose library of, it is said, 80,000 volumes would be useful in preparing his inaugural address. He then appears to have established himself in a modest house in the vicinity of the Church of San Giustina, on the Prato della Valle (now Piazza Vittorio Emmanuele). Little did he then think as he strolled about the Prato, then a Grand Place for fêtes and spectacles of all kinds, that his statue would be one of seventy-seven which would be erected in years long after to commemorate great men who had made Padua illustrious for all time.

DURING THE FIRST few years at Padua, Galileo displayed an extraordinary and versatile activity. He wrote a number of treatises, chiefly for the use of his pupils, among the larger of which may be mentioned: On Military Architecture, on Fortifications, on Mechanics, on the Sphere, on Accelerated Motion, on Gnomonics. All of these attained a wide circulation in manuscript copies; some were not printed until long afterwards, while others, like the paper on Gnomonics, are unfortunately lost. Others, again, strayed beyond the pupils and friends for whom they were intended, and found their way into the hands of persons who did not scruple to claim and publish them as their own.

Galileo's treatise on Mechanics ("Della Scienza Meccanica"), written in 1594, deals with the powers of the lever, pulley, and screw, and concludes with an account of the Archimedian Screw for raising water, followed by a short fragment "On the Force of Percussion."

The "Treatise on the Sphere," which was first published at Rome in 1656 (fourteen years after Galileo's death), is supposed by some authors to be apocryphal, as it teaches the Ptolemaic cosmogony, placing the earth immovable in the centre, and adducing the usual arguments. But this does not make the work necessarily apocryphal, for we have it under his own hand that for some years he taught the Ptolemaic system in his classes out of compliance with popular feeling, although at heart he was a follower of Copernicus.

In his first letter to Kepler, dated 4th August 1597, and acknowledging the receipt of the latter's *Mysterium Cosmographicum,* he says:

> I have as yet read nothing beyond the preface of your book, from which, however, I catch a glimpse of your meaning, and feel great joy on meeting with so powerful an associate in the pursuit of truth, and, consequently, such a friend to truth itself; for it is deplorable that there should be so few who care about truth, and who do not persist in their perverse mode of philosophising. But as this is not the fit time for lamenting the melancholy condition of our times, but for congratulating you on your elegant discoveries in confirmation of the truth, I shall only add a promise to peruse your book dispassionately, and with the conviction that I shall find in it much to admire.

> This I shall do the more willingly because many years ago I became a convert to the opinions of Copernicus, and by his theory have succeeded in explaining many phenomena which on the contrary hypothesis are altogether inexplicable. I have arranged many arguments and confutations of the opposite opinions, which, however, I have not yet dared to publish, fearing the fate of our master, Copernicus, who, although he has earned immortal fame among a few, yet by an infinite number (for so only can the number of fools be measured) is hissed and derided. If there were many such as you I would venture to publish my speculations, but since that is not so I shall take time to consider of it.

In the early summer of 1593 Galileo contracted an illness, which nearly proved fatal, and from which he suffered at intervals all through life. A party of three young men (so Viviani tells the story), of whom Galileo was one, "were enjoying at an open window in the country a current of air which was artificially cooled by a fall of water. They, unfortunately, fell asleep under its influence, and so great was its effect on Galileo's hitherto robust constitution that a severe chronic disorder ensued, which showed itself in acute pains in the body, accompanied by frequent haemorrhages and loss of appetite and sleep. The others suffered still more severely, for one died in a few days, while the third became deaf and died in a short time after."

WHILE CARRYING on his professorial duties, giving private lessons, and writing learned tracts, Galileo was occupying himself profitably in other ways. Thus it would appear that his skill as an engineer was often in request by the State, for which he is said to have designed, or superintended the construction of, various works and machines. Amongst the latter may here be noted a machine for raising water, of small dimensions but of great power, so that one horse could raise the water and distribute it through twenty channels. For this the Venetian republic gave him the exclusive use for twenty years, under a patent dated 15th September 1594; but, although it was tried with success in the garden of the Contarini Palace in Venice, it never came into much use.

About 1596–7 he invented a more profitable instrument, his Geometrical and Military Compass. This useful instrument is now called the Sector, and is to be found in most cases of mathematical instruments. It consists of two straight rulers connected by a joint so that they can be set to any required angle. On one face are four pairs of lines.

- Arithmetical lines, which serve for the division of lines, the solution of the Rule of Three, the equalisation of money, the calculation of interest.
- Geometrical lines, for reducing proportionally superficial figures, extracting the square root, regulating the front and flank formations of armies, and finding the mean proportional.
- Stereometrical lines, for the proportional reduction of similar solids, the extraction of the cube root, the finding of two mean proportionals, and for the transformation of a parallelepiped into a cube.
- Metallic lines, for finding the proportional weights of metals, and other substances, for transforming a given body into one of another material and of a given weight.

On the other side of the instrument are:

- Polygraphic lines, for describing regular polygons, and dividing the circumference into equal parts.
- Tetragonal lines, for squaring the circle or any other regular figure, for reducing several regular figures to one

figure, and for transforming an irregular rectilineal figure into a regular one.
- JOINED LINES, used in the squaring of the various portions of the circle and of other figures contained by parts of the circumference, or by straight and curved lines together.

There is joined to the compass a quadrant, which, besides the usual divisions of the astronomical compass, has engraved on it a squadron of bombardiers, and, in addition, transversal lines, used for taking the inclination of a scarp of a wall.

From the encouraging reception given to this invention, orders for which came from all parts of Europe, Galileo determined to open a workshop in his own house, no doubt in order that the manufacture of this and other Scientific apparatus might proceed uninterruptedly under his personal supervision, and so be less liable to piracy. This we gather from an entry in his Day-book under 5th July 1599.

> Messer Marcantonio Mazzoleni, mechanician, comes to reside in my house to work for me, and at my cost, on mathematical instruments, I undertaking to bear the expenses of him, his wife, and child, and to pay him in addition the sum of six ducats *per annum*.

This is one of the instances, before referred to, where Galileo's manuscripts and ideas got into the hands of people who were not ashamed to publish them as their own, and to denounce the author as an impudent plagiarist. Some years after the invention, finding that his right to it was being disputed, Galileo published a description of it and dedicated the book to his pupil, Prince Cosimo, son of the Grand Duke of Tuscany. This was speedily followed by another book, the production of Baldassare Capra, a young Milanese, in which he claimed the invention as his own, and accused Galileo of piracy. The matter was brought before the University authorities, and Galileo was able to show that he had made the invention as early as 1597, and had explained its construction and uses to numerous persons in Padua and elsewhere, amongst them being Capra himself! These statements were supported by depositions of well-known men, as Gio. Fran. Sagredo, Giacomo Badovere, Mazzoleni (Galileo's mechanician), Giacomo Alvise Carnaro, and Fra Paolo Sarpi. The last-mentioned only need be quoted. Writing from Venice under date 20th April 1607, he affirms and attests that he had carefully compared the two works in dispute, and had found Capra's to be little more than a Latin translation of Galileo's Italian. He then goes on:

> I further affirm that about ten years ago in Padua Signor Galileo showed me the instrument (described in his book) and explained its uses; and that about two years later, the said Signor made me a present of one, which I still have in my possession.

As Capra made no defence, the University authorities did not take long to decide. Under date 4th May 1607, they unanimously decreed that his book was a scandalous plagiary and an insult to Galileo and the University (of which Capra was a member); that all the copies in the possession of the author and his publisher, to the number of 483, be given up to them for suppression "in the way that seems best"; and that proceedings be taken against the printer and publisher.

Galileo afterwards published a full account of this affair in his "Difesa contro alle Calunnie ed Imposture di Baldassare Capra." The first part is taken up with a defence of his views on the new star of 1604, which Capra had attacked in another publication—moved thereto, as Viviani says, "by envy of the universal applause which accrued to Galileo from his lectures on the subject."

From the opening passages of Galileo's "Il Saggiatore" (1623), it would appear that the real author of the book which went under

Capra's name was Simon Mayer, a German graduate of Padua, whom we shall meet with later on arrogating to himself the merit of two of Galileo's astronomical discoveries. On this occasion, as soon as he found our philosopher intent on resenting the injury, he hastily quitted Italy, leaving his friend Capra to bear alone the shame of the exposure which followed.

To THIS PERIOD (1602) may be referred Galileo's invention of the air thermometer. The date is uncertain, for while Viviani asserts that the instrument was designed during the first term of his Professorship at Padua (1592-98), other evidence, on which Galileo rested his claim when contested some years later, will only carry us back to about 1602, Thus, Castelli, writing to Ferdinando Cesarini, 20th September 1638, says:

> I remember an experiment which our Signor Galileo had shown me more than thirty-five years ago. He took a small glass bottle about the size of a hen's egg, the neck of which was two palms long (about 22 inches), and as narrow as a straw. Having well heated the bulb in his hand, he inserted its mouth in a vessel containing a little water, and, withdrawing the heat of his hand from the bulb, instantly the water rose in the neck more than a palm above its level in the vessel. It is thus that he constructed an instrument for measuring the degrees of heat and cold.

Galileo's friend, Sagredo, was the first to divide the tube into 100 degrees in 1613. He also appears to have experimented with closed tubes from about 1615 ; but it was not until many years after (1653) that the practice of hermetically closing the orifice after exhausting the air was introduced. The credit of this capital improvement is due to

Leopoldo de Medici, brother of Ferdinando II, who adopted the plan of expelling the air by boiling the spirit and sealing the end of the tube whilst the contained liquid was in an expanded state, thus depriving the instrument of its barometrical character and making it a true thermometer.

We have said above that Galileo's right to this invention was contested, as in the case of his geometrical and military compass, and in others to which we shall come later on. It was claimed for Porta, Santorio, and Paolo Sarpi, in Italy; for Robert Fludd and Francis Bacon in England; and for the Dutchman, Drebbel. But in disproof of these claims it will be enough to say that the first mention of the instrument by Porta occurs in 1606, by Santorio in 1612, by Sarpi and Fludd in 1617, by Bacon in 1620, and by Drebbel in 1621.

~~JESUS HEALING THE SICK~~

GALILEO SOLVING THE WORLD'S PROBLEMS WITH TECHNOLOGY

And great multitudes came unto him, having with them those that were disoriented, chronically late, cold, easily swindled, and many others, and cast them down at Galileo's feet; and he remedied their troubles.

IN 1604 THE attention of astronomers was attracted to a new star which suddenly appeared with great splendour in the constellation Serpentarius. Maestlin, who was one of the first to notice it, thus describes it:

> How wonderful is this new star! I am certain that I did not see it before 29th September, nor indeed, on account of several cloudy nights, had I a good view till 6th October. Now that it is on the other side of the sun, instead of surpassing Jupiter as it did, and almost rivalling Venus, it scarcely equals the Cor Leonis, and hardly surpasses Saturn. It continues, however, to shine with the same bright and strongly sparkling light, and changes colour almost every moment, now tawny, then yellow, presently purple, and red, and, when it has risen above the vapours, most frequently white.

Galileo appears to have noticed the new star very soon after Maestlin, i.e. on 10th October; and (whether by accident or design is not known) he chose for the subject of his ordinary lectures in the session then opening the theory of the planets. This afforded his auditors the wished-for opportunity of getting his views on the new phenomenon, the appearance of which had given rise to the most bewildering statements. Some said it was a light in the inferior regions of space—"the elementary sphere," but they did not explain how it got there; others, that it was an old star hitherto unnoticed; others again, founding their opinion on abstruse teleological grounds, declared that new stars were created by God from time to time, and that this was one of them; while, to add to the confusion, the astrologers deduced from it the wildest forebodings.

After carefully watching the star for some time (it lasted eighteen months), Galileo resolved to expound his views in three extra-ordinary lectures, which were delivered to the public in the great hall *(Aula Magna)* of the University early in January 1605. In the opening sentences of the first lecture, which are the only parts preserved to us, he took occasion to rebuke his auditors for their general insensibility to the wonders of creation daily exposed to their view, in no way less admirable than the new prodigy, to hear an explanation of which they had hurried in crowds to his lecture room. As regards the star itself we know, from references in his other writings, that he demonstrated that it was neither a meteor, nor yet a body existing from all time, and only now noticed, but a body which had recently appeared and would again vanish. Unlike his contemporaries, Tycho Brahe and Kepler, who thought that new stars (and comets) were temporary conglomerations of a cosmical vapour filling space; or, as is now thought, the result of some catastrophe or collision whereby immense masses of incandescent gases are produced, Galileo suggested that they might be products of terrestrial exhalations of extreme tenuity, at immense distances from the earth, and reflecting the sun's rays—an hypothesis which, as we shall see later on, he also applied to comets. From the absence of parallax he showed that the new star could not be, as the current theory held, a mere meteor engendered in our atmosphere, and nearer to us than the moon, but that it must be situated among the most remote heavenly bodies.

This was inconceivable to the Aristotelians, whose notions of a perfect and unchangeable heaven, subject neither to growth nor to decay, were quite at variance with the introduction of any such new body. It is hard to say whether Galileo's colleagues, bred in the old philosophy, were more annoyed at the appearance of the star, or at his calling attention to it so publicly and forcibly. Controversy was now unavoidable at Padua, as a few years before at Pisa, and Galileo did not shirk it. He boldly threw down the gauntlet in favour of the Copernican theory, and repudiated the old systems of Aristotle and Ptolemy, which up to that time he had taught in his classes.

The Aristotelians put forth one of their best advocates, Antonio da Montepulciano, to confute Galileo's views, and the latter replied in the only way possible—by ridicule. With the aid of some of his pupils he wrote and printed an exquisite squib in the Paduan dialect, entitled, "Dialogo de Cecco di Ronchitti da Bruzene in Perpuosito del la Stella Nuova."

3

GALILEO, PROFESSOR IN PADUA (CONTINUED)

1592–1610

In September 1598, the first period of Galileo's appointment expired, which the Venetian Senate were in no hurry to renew formally; not because of any doubt as to the incumbent's fitness, but because of the dreaded increase of salary which was usually expected on re-appointment, and which all governments like to evade if they can. Galileo himself had allowed the first term to expire, and had nearly completed one year of the second, before taking any steps in the matter. Then, about the middle of 1599, with the advice and assistance of Pinelli, Gianfrancesco Sagredo, a young Venetian of great promise, and other friends, he formally requested that his appointment be renewed, on an increased salary, as a precedent for which he cited the case of Professor Magini at Bologna, who on re-appointment was awarded a salary much in excess of that enjoyed by Galileo at Padua.

The question of augmentation was apparently long and stoutly opposed on one side, and as hotly pressed on the other. The Doge, Contarini, complained that he was pestered on the subject, not only by Galileo's friends, but by his own household. "If Galileo," he cried, "is not content with his salary he can resign." Moletti, he pointed out, had never more than 300 florins, and it was an understood thing that the

Professors should eke out their incomes by private lessons. In the end the appointment was renewed, on 28th October 1599, for a further period of six years, commencing 27th September 1598. The salary was fixed at 320 florins (about £70), and Sagredo was grumblingly bidden to warn his friend not to expect any further augmentation, "as the Senate did not choose to make his case a precedent for every learned and hungry foreigner who might think fit to press a similar claim."

Galileo's reputation as a teacher was now widely spread over Europe, and numbers of young foreigners flocked to Padua to attend his lectures. Amongst these are noted the Archduke Ferdinand, afterwards Emperor of Germany; the Landgrave of Hesse; the Princes of Alsace and Mantua; and a Prince Gustavus of Sweden, often confounded with the great Gustavus Adolphus. Our own William Harvey, the discoverer of the circulation of the blood, was a student at Padua, 1598–1602, and would no doubt be a frequent attendant.

IT IS NOT certain when Galileo moved from his modest house near the Church of San Giustina; but towards the end of 1599 we find him installed in a large house in the Via Vignali, now Via Galileo. Here he was able to accommodate suitably his private pupils, of whom he now had large numbers resident with him, both Italians and foreigners. His private papers show that he looked after the catering himself, which must have been no light matter, seeing that he often had as many as twenty boarders at a time, some of whom were accompanied by servants or followers of some kind. Inclined to be prodigal by nature, Galileo made nothing out of his pupils' keep, being content if their contributions covered the expenses of house-keeping.

With the house went a garden, to which in 1603 he added an adjoining piece of ground containing a large arbour and vine-trellises. The care of this garden, the flowers, the fruits, the vines, was his great delight. He saw to everything and did much himself in the way of weeding, pruning, tying. Indeed, all his life he was very fond of gardening, which he followed as much as an education as a recreation and distraction from severer studies. Here he took his pleasure, sometimes alone, revolving deep things in philosophy, but more often surrounded by groups of his friends, or pupils, whom he now sported with, now charmed with performances on the lute, which he touched with a master's hand—

As sweet and musical
As bright Apollo's lute, strung with his hair.

They conversed, as they strolled along the arbour and under the spreading vines, on any subject that presented itself—the germination of seeds, the nutrition and vegetation of plants, the making of wines, philosophising all the while "from grave to gay, from lively to severe." We can imagine how charming and instructive his conversations were, for Galileo had the habit of allowing no natural phenomena, however trivial in appearance, to escape him, and he found the advantage of this in his lectures and writings, as it furnished him with a stock of homely illustrations to which the daily experience of his hearers and readers readily assented, and which he could show to be identical in principle with the matter in hand.

PARABLE OF THE CICADA

Galileo told his students many things in parables, saying:

"Once upon a time, in a very lonely place, there lived a man endowed by nature with extraordinary curiosity and a very penetrating mind. For a pastime he raised birds, whose songs he much enjoyed; and he observed with great admiration the happy contrivance by which they could transform at will the very air they breathed into a variety of sweet songs.

"One night this man chanced to hear a delicate song close to his house, and being unable to connect it with anything but some small bird he set out to capture it. When he arrived at a road he found a shepherd boy who was blowing into a kind of hollow stick while moving his fingers about on the wood, thus drawing from it a variety of notes similar to those of a bird, though by quite a different method. Puzzled, but impelled by his natural curiosity, he gave the boy a calf in exchange for this flute and returned to solitude. But realising that if he had not chanced to meet the boy he would never have learned of the existence of a new method of forming musical notes and the sweetest songs, he decided to travel to distant places in the hope of meeting with some new adventure.

"The very next day he happened to pass by a small hut within which he heard similar tones; and in order to see whether this was a flute or a bird he went inside. There he found a small boy who was holding a bow in his right hand and sawing upon some fibres stretched over a hollowed piece of wood. The left hand supported the instrument, and the fingers of the boy were moving so that he drew from this a variety of notes, and most melodious ones too, without any blowing. Now you who participate in this man's thoughts and share his curiosity may judge of his astonishment. Yet finding himself now to have two unanticipated ways of producing notes and melodies, he began to perceive that still others might exist.

"His amazement was increased when upon entering a temple he heard a sound, and upon looking behind the gates discovered that this had come from the hinges and fastenings as he opened it. Another time, led by curiosity, he entered an inn expecting to see someone lightly bowing the strings of a violin, and instead he saw a man rubbing his fingertip around the rim of a goblet and drawing forth a pleasant tone from that. Then he observed that wasps, mosquitoes, and flies do not form single notes by breathing, as did the birds, but produce their steady sounds by swift beating of their wings. And as his wonder grew, his conviction proportionately diminished that he knew how sounds were produced; nor would all his previous experiences have sufficed to teach him or even allow him to believe that crickets derive their sweet and sonorous shrilling by scraping their wings together, particularly as they cannot fly at all.

"Well, after this man had come to believe that no more ways of forming tones could possibly exist—after having observed, in addition to all the things already mentioned, a variety of organs, trumpets, fifes, stringed instruments, and even that little tongue of iron which is placed between the teeth and which makes strange use of the oral cavity for sounding box and of the breath for vehicle of sound—when, I say, this man believed he had seen everything, he suddenly found himself once more plunged deeper into ignorance and bafflement than ever. For having captured in his hands a cicada, he failed to diminish its strident noise either by closing its mouth or stopping its wings, yet he could not see it move the scales that covered its body, or any other thing. At last he lifted up the armour of its chest and there he saw some thin hard ligaments beneath; thinking the sound might come from their vibration, he decided to break them in order to silence it. But nothing happened until his needle drove too deep, and transfixing the creature he took away its life with its voice, so that he was still unable to determine whether the song had originated in those ligaments. And by this experience his knowledge was reduced to diffidence, so that when asked how sounds were created he used to answer tolerantly that although he knew a few ways, he was sure that many more existed which were not only unknown but unimaginable.

"I could illustrate with many more examples Nature's bounty in producing her effects, as she employs means we could never think of without our senses and our experiences to teach them to us—and sometimes even these are insufficient to remedy our lack of understanding. So I should not be condemned for being unable to determine precisely the way in which comets are produced, especially in view of the fact that I have never boasted that I could do this, knowing that they may originate in some manner that is far beyond our power of imagination. The difficulty of comprehending how the cicada forms its song even when we have it singing to us right in our hands ought to be more than enough to excuse us for not knowing how comets are formed at such immense distances. Let us therefore go no further than our original intention, which was to set forth the questions that appeared to upset the old theories, and to propose a few new ideas."

AMONG THE CROWD of noble and learned men, with whom at this

period Galileo had cemented a friendship which was only to be severed by death, may be mentioned (besides Pinelli) Fra Paolo Sarpi of the Order of Servites, Theologian and Counsellor of the Venetian Republic, and afterwards known as the Machiavelli of Venice; Fra Fulgenzio Micanzio, Sarpi's devoted friend and colleague; Fabrizio d'Acquapendente, the famous surgeon of Padua, who has been called "the Columbus of the human body"; Antonio de Medici; General del Monte; and young Gianfrancesco Sagredo, who developed into "a witty and eccentric patrician, whose house at Venice resembled Noah's ark, containing all manner of beasts." Pinelli took an opportunity of his correspondence with Tycho Brahe to recommend Galileo "as a man whose friendship would be worth while cultivating." Tycho addressed a letter to Galileo, under date 4th May 1600, but the acquaintance went no farther, and the great Dane died at Prague, 13th October 1601.

IN 1604, when the second term of Galileo's Professorship expired, he applied for its renewal with a further increase of salary; but, as usual, the Venetian authorities were slow to move. At this time Vincenzio Gonzaga, the reigning Duke of Mantua, was anxious to attach him to his court and person, and made some tempting overtures, but nothing came of them. However, while in Florence during the summer of 1605, Galileo took an opportunity of interesting his young pupil Cosimo, son of the Grand Duke, in his case; and at length, and mainly through the influence of the Tuscan Ambassador at Venice (prompted by Cosimo), he was reappointed for a third term of six years, with an augmented salary of 520 florins (about £115), by decree of 5th August 1606.

His public lectures were at this time so thronged that the ordinary classrooms, large as they were, were often insufficient to contain his audiences. His more popular lectures, as, for instance, those on the new star, were delivered in the Aula Magna, the great hall of the University, and capable of holding 1,000 persons. Even this, according to Drinkwater, was not large enough, for "on several occasions he was obliged to adjourn to the open air."

LECTURE IN THE EUGANEAN HILLS

Now when Galileo saw the crowds of students, he went up into the Euganean hills and sat down. His students came to him, and he began to teach them.

He said:

THE PROSPERITUDES

"Prosperous are the strong in Progress, for theirs is the coming Utopia.

"Prosperous are those who analyse, for they will reach conclusions.

"Prosperous are the technologically inclined, for they will conquer the universe.

"Prosperous are those who hunger and thirst for knowledge, for they will be satiated.

"Prosperous are the skeptical, for they will be shown empirical evidence.

"Prosperous are the pure of mind, for they will know Science.

"Prosperous are the experimentalists, for they will be called students of Science.

"Prosperous are those who are persecuted because of their rationality, for theirs is the coming Utopia.

"Prosperous are you when people question you, argue with you, and falsely say all kinds of ignorant things against you because of Science. Persevere and press onward, because great is your progress toward Utopia, for in the same way they persecuted the Scientists who were before you."

PUBLICATION

"You may consider remaining perfectly silent about your work in order to save yourself any occasion for being the unhappy target of such sharpshooting. But if the Scientist abjures his Science, how can he be called Scientific? I myself have certainly not lacked opportunities to put forth other works that would perhaps be no less astonishing to the schools of philosophy and no less important to Science than those published previously. But to protect myself from ill feeling and stubborn opposition, I contented myself merely with the opinion and judgment of a few gentlemen, my real friends, to whom I communicated my thoughts. In discussions with these men I have enjoyed that pleasure which accompanies the opportunity to impart what one's mind brings forth bit by bit, and at the same time I avoided any renewal of those stings which I had previously experienced with so much vexation. Demonstrating in no small degree their approval of my ideas, these gentlemen have managed for a variety of reasons to draw me away from the resolution I had made. In the same way, let your Scientific accomplishments be published, so that others may see your good works and properly acknowledge the superiority of Logic."

AUTHORITY

"You say you do not wish to be numbered among those who affront the sages by disbelieving and contradicting them. I say I do not wish to be counted as an ignoramus and an ingrate toward Nature and toward Science; for if they have given me my senses and my reason, why should I defer such great gifts to the errors of some man? Why should I believe blindly and stupidly what I wish to believe, and subject the freedom of my intellect to someone else who is just as liable to error as I am? For I tell you that unless your knowledge surpasses that of the Peripatetics and teachers of the Ptolemaic system, you certainly will not arrive at the coming Utopia."

PERSECUTION

"You have heard it said that you should not be upset by obstinate attacks, since in the end those will rebound upon their authors and merely render your own reasoning more lively and attractive, furnishing as they do clear proof that your essays are of an uncommon nature. They have pointed out to you the familiar maxim that vulgarity and mediocrity receive little or no attention and are soon left in the cold, while men's minds turn to the revelation of wonders and transcendent things—though these indeed may give rise in ill-tempered minds to envy, and thereby to slander.

"But I tell you, it is useless to hold your peace, because those who are so anxious to make trouble for you now have recourse to attributing to you the works of others. In this way they stir up bitter fights against you, something that I believe never happens without indicating some insane passion. Therefore, do your best to see that these acts do not escape notice, and to discourage those who refuse to let sleeping dogs lie and who stir up trouble with men that are at peace.

"How many men have attacked my writings, and under what disguises! The material contained therein ought to have opened to the mind's eye much room for admirable speculation; instead it met with scorn and derision. Many people disbelieved it or failed to appreciate it. Others, not wanting to agree with my ideas, advanced ridiculous and impossible opinions against me; and some, overwhelmed and convinced by my arguments, attempted to rob me of that glory which was mine, pretending not to have seen my writings and trying to represent themselves as the original discoverers of these impressive marvels."

POETRY AND SCIENCE

"It has been suggested by a man called Sarsi that our intellects should be enslaved to the poets and philosophers who came before. But I tell you, my mind will not be satisfied nor set at rest by a little poetic flower that is not followed by any fruit at all. Nature takes no delight in poetry. That is a very true statement, even though Sarsi appears to disbelieve it and acts as if acquainted with neither nature nor poetry. He seems not to know that fables and fictions are in a way essential to poetry, which could not exist without them, while any sort of falsehood is so abhorrent to nature that it is as absent there as darkness is in light. Therefore, anyone who studies poetry, except as an idle pastime, forsakes his calling as a Scientist, and anyone who tries to become a poet commits blasphemy."

IGNORANCE

"Again, you have heard it said by Sarsi that I claimed my opinion was certain to be swiftly carried by the winds to Rome. But I tell you, that usually happens only with the words of great and celebrated men, which really far exceeds the bounds of my ambition. It is true, though, that in reading Sarsi's book I have wondered that what I said never did reach Sarsi's ears. Is it not astonishing that so many things have been reported to him which I never said, nor even thought, while not a single syllable reached him of other things that I have said over and over again? But perhaps the winds that blow the clouds and those chimeras and monsters that tumultuously take shape in them had not the strength to carry solid and weighty things. Therefore, do not attempt to refute arguments you do not understand, for you cannot make even one statement that is valid or meaningful. All you need to say is simply, 'This bears further investigation'; anything beyond this comes from the ignorant one."

OBSERVATION

"You have heard it said by Sarsi that the sense of sight deserves our utmost confidence, and it is impossible for us to be deceived by a spurious object whenever that may be set beside a real one. But I tell you, I do not possess such a perfect faculty of discrimination. I am more like the monkey that firmly believed he saw another monkey in a mirror, and the image seemed so real and alive to him that he discovered his error only after running behind the glass several times to catch the other monkey. Therefore, if an experiment requires you to make one observation, you must make the observation at least twice. See what is there to be seen, and do not turn away from the evidence of your own eyes."

REASONING

"You have heard it said by Sarsi, following authors from the past, that the Babylonians cooked their eggs by whirling them in slings. But I tell you, if

Sarsi wishes for me to assent to this statement, I must say that the cause of this effect was very different from what he suggests. To discover the true cause I reason as follows: "If we do not achieve an effect which others formerly achieved, then it must be that in our operations we lack something that produced their success. And if there is just one single thing we lack, then that alone can be the true cause. Now we do not lack eggs, nor slings, nor sturdy fellows to whirl them; yet our eggs do not cook, but merely cool down faster if they happen to be hot. And since nothing is lacking to us except being Babylonians, then being Babylonians is the cause of the hardening of eggs, and not friction of the air." And this is what I wished to discover. Is it possible that Sarsi has never observed the coolness produced on his face by the continual change of air when he is riding post? If he has, then how can he prefer to believe things related by other men as having happened two thousand years ago in Babylon rather than present events which he himself experiences? This is a grave error in reasoning which is unbecoming of a true Scientist. Be perfectly logical, therefore, as Science which rules the cosmos is perfect."

GIVING TO THE IGNORANT

"Be careful not to downplay your rationality in front of others so that you do not upset them. If you do, you will not be furthering the cause of Science.

"So when you give knowledge to the ignorant, do not write it in private letters, as the lesser Scientists do in the courts and in the royal societies, to be discovered later by posterity. Truly I tell you, they have received their recognition in full. But when you give knowledge to the ignorant, publish it in books under your own name, so that your rationality may be made evident. Then Science, which thrives on the open exchange of ideas, will make you prosperous."

PETITIONING

"And when you seek to increase your own knowledge, do not go to the popular scientists, for they love to pontificate standing in the auditoria and on the lecture circuit to be seen by others. Truly I tell you, publicity is their only reward. But when you seek knowledge, go into your laboratory, close the door and petition to Science, which governs the world, both seen and unseen. Then Science, which reveals the hidden things, will grant you knowledge. And when you petition, do not keep on babbling like Aristotelians and Ptolemaics, for they think they will arrive at knowledge because of their many words. Do not be like them, for Science speaks in numbers rather than in words.

"This, then, is how you should petition for knowledge:

O Science, Which orders the Universe:
Hallowed be Thy precepts.
Thy supremacy be undisputed,
Thy validity be recognised,
On Earth as it is throughout the Cosmos.
Give us this day our empirical evidence.
And correct our miscalculations,
As we correct those who miscalculate around us.
And let us not slip into human error,
But deliver us from ignorance.

"For if you correct other people when they miscalculate, Science will also correct you. But if you do not correct their miscalculations, Science will not correct your miscalculations."

DISPUTATION

"When you dispute, do not reply to arguments with additional limitations, distinctions, logical technicalities, philosophical jargon, and other idle words. Truly I tell you, in sustaining one error you will commit a hundred others that are more serious, and produce always greater follies in your camp. But when you dispute, put forward only logical arguments and empirical evidence, so that it will be obvious to others that you are disputing rationally. Then Science, which advances by means of intellectual debate, will make you prosperous.

"Do not dispute with me, as I am outside the circle of those ancient and modern authors against whom your other teachers are contending. I mean only to address those men, ancient or modern, who try in all their studies to investigate some truth in nature. I mean to steer clear of those who ostentatiously engage in noisy contests merely to be popularly judged victors over others and pompously praised."

TWO TEACHERS

"No one can study under two teachers. Either you will mistrust the one and believe the other, or you will be devoted to the one and skeptical of the other. You cannot study both Science and ancient philosophy."

EMPIRICISM

"Therefore I tell you, do not put the testimony of writers ahead of what experience shows you. To adduce more witnesses serves no purpose—so far as authority is concerned, yours alone is as effective as an army's in rendering events true or false. You take your stand on the authority of many poets against our experiments. I reply that if those poets could be present at our experiments they would change their views, and without disgrace they could say they had been writing hyperbolically—or even admit they had been wrong.

"Well, if we cannot have the presence of your poets (who, as I say, would yield to experience), we do have at hand archers and catapultists, and you may see for yourself whether citing your authorities to them can strengthen their arms to such an extent that the arrows they shoot and the lead balls they hurl will take fire and melt in the air. In that way you will be able to find out just how much force human authority has upon the facts of Nature, which remains deaf and inexorable to our wishes.

"Do not persist in trying to prove by means of witnesses something that you may see for yourself at any time by means of experiment. Witnesses are examined in doubtful matters which are past and transient, not in those which are actual and present. A judge must seek by means of witnesses to determine whether Peter injured John last night, but not whether John was injured, since the judge can see that for himself. But even in conclusions which can be known only by reasoning, I say that the testimony of many has little more value than that of few, since the number of people who reason well in complicated matters is much smaller than that of those who reason badly. If reasoning were like hauling I should agree that several reasoners would be worth more than one, just as several horses can haul more sacks of grain than one can. But reasoning is like racing, and not like hauling, and a single Arabian steed can outrun a hundred plow-horses. So when you bring in a multitude of authors, instead of strengthening your conclusion you merely ennoble my case by showing that I have out-reasoned many men of great reputation."

ASK, SEEK, READ

"Ask for knowledge and it will be given unto you; seek knowledge and you will find it; read and the book will be opened unto you. For everyone who asks receives, the one who seeks finds, and for the one who reads, the book will be opened.

"Philosophy is written in this grand book, the universe, which stands continually open to our gaze. But the book cannot be understood unless one first learns to comprehend the language and read the letters in which it is composed. It is written in the language of mathematics, and its characters are triangles, circles, and other geometric figures without which it is humanly impossible to understand a single word of it; without these, one wanders about in a dark labyrinth. So in everything, look for underlying patterns and principles, for these sum up the Laws of Nature and Mathematics."

THE NARROW WALLS

"Do not think you can enclose the host of good philosophers within four walls. For they fly, and they fly alone, like eagles, and not in flocks like starlings. It is true that because eagles are rare birds they are little seen and less heard, while birds that fly like starlings fill the sky with shrieks and cries, and wherever they settle befoul the earth beneath them. Yet if true philosophers are like eagles, they are not unique like the phoenix. The crowd of fools who know nothing is infinite. Those who know very little of philosophy are numerous. Few indeed are they who really know some part of it, and only Science knows all."

TRUE AND FALSE DISCIPLES

"Judge not a man's philosophical opinions by the number of his followers. Though the number of disciples of the best philosophers may be quite small, I do not conclude conversely that those opinions and doctrines are necessarily perfect which have few followers, for I know well enough that some men hold opinions so erroneous as to be rejected by everyone else. Therefore, you must study each author's works sufficiently to judge their merit. A good philosopher cannot write bad books, and a bad philosopher cannot write good books. Every philosopher that does not write good books shall eventually pass into obscurity. Thus, by their works you shall judge them.

"To put aside hints and speak plainly, dealing with Science as a method of demonstration and reasoning capable of human pursuit, I tell you that the more this partakes of perfection the smaller the number of propositions it will promise to teach, and fewer yet will it conclusively prove. Consequently the more perfect it is the less attractive it will be, and the fewer its followers. On the other hand, magnificent titles and many grandiose promises attract the natural curiosity of men and hold them forever involved in fallacies and chimeras, without ever offering them one single sample of that sharpness of true proof by which the taste may be awakened to know how insipid is the ordinary fare of philosophy. Such things will keep an infinite number of men occupied, and that man will indeed be fortunate who, led by some unusual inner light, can turn from dark and confused labyrinths in which he might have gone perpetually winding with the crowd and becoming ever more entangled.

"Long experience has taught me this about the status of mankind with regard to matters requiring thought: the less people know and understand about them, the more positively they attempt to argue concerning them, while on the other hand to know and understand a multitude of things renders men cautious in passing judgment upon anything new."

When Galileo had finished saying these things, the crowds were amazed at his teaching, because he taught as one who had knowledge, and not as their teachers of Aristotelian physics.

~~THE SERMON ON THE MOUNT~~

THE LECTURE IN THE EUGANEAN HILLS

Consider the pendulums of the air, how they move; they slacken not, neither do they hasten: And yet I say unto you, that even DaVinci in all his genius could not comprehend the motion of these.

HAVING BROUGHT our account of Galileo's public career up to the eve (1609) of his immortal discoveries in astronomy, we must, before proceeding, retrace our steps and give some account of his private life.

We have seen that by the death of his father, in 1591, he had become the head of the family. This position, always attaching a grave responsibility to its possessor, was at this time, and particularly in the present case, fraught with much anxiety. Not only had Galileo, on his slender resources, to keep himself in Padua and provide for the requirements of the household in Florence, but it was his duty to see to his brother's setting out in life, and a still more sacred duty, to find a suitable husband for his unmarried sister, Livia. In Italy in those days, a girl's education being finished, two paths were open—not always for her to choose. One led to the cloister, the other to the house of a husband. It had, apparently, been the family intention for Livia to take the veil, but so great was her aversion to a convent life, that her brother did not insist, much as the arrangement would have saved him in trouble and expense.

Though his sister, Virginia, had been married to Benedetto Landucci before the father's death, the burden of providing the dowry had fallen on Galileo, who, pressed on all sides for money, had been unable immediately to pay the amount. At length, in May 1593, not choosing to wait any longer, and not caring who went short so long as he got his due, Landucci threatened to proceed to harsh measures—in fact, to have Galileo arrested for debt the next time he set foot in Florence!

As if the worries and burdens of his father's family were not enough, Galileo must needs add to them the cares of a family of his own. In 1599, he entered into amorous relations with a Venetian lady, Marina Gamba, by whom he had three children, Virginia, born 13th August 1600; Livia, born 18th August 1601; and Vincenzio, born 21st August 1606.

4

GALILEO, PROFESSOR IN PADUA (CONTINUED)

1592–1610

In 1607 Galileo had been studying the *De Magnete* of Dr William Gilbert of Colchester, a book which had for him always a great fascination, and for two reasons—first, its arguments traversed many of the principles of the Aristotelian School; and second, it contained a number of original experiments in electricity and magnetism, coupled with philosophical reflections of a far-reaching kind, which appealed to his own daring spirit.

Doubtless owing to his other occupations, Galileo's researches in magnetism did not take him far. The main results are given in two letters to Curzio Picchena, Chief Secretary to the Grand Duke of Tuscany, dated 16th November and 9th December 1607. From these we learn that he had made many experiments with loadstones, which resulted in his devising an armature by which the portative force of a stone could be considerably increased. He observed that the longer a stone sustains a weight, the more it gains in strength. He also found that smaller stones were usually more powerful than larger ones. Thus, a small stone which he had picked up in Venice was much more powerful than one of 5 lbs. belonging to his friend, Sagredo. This latter could normally sustain a weight of 6½ lbs, but when provided with Galileo's armature it was capable of supporting more than 12 lbs. By breaking up large stones and shaping the pieces in a certain way he was able to make each piece sustain thirty to forty times the weight borne by the original stone. The Abbé Castelli tells us in his treatise on the magnet (c. 1639–40):

> I have seen such a loadstone, only 6 oz. in weight, armed with iron by the untiring industry of Signor Galileo, and presented to his Highness the Grand Duke Ferdinando, which lifts 15 lbs. of iron worked into the shape of a sepulchre.

Gilbert was one of the first persons who arrived at general, though confused, notions on the subject of gravitation. In his *De Magnete* he explains the influence of the earth upon the moon by comparing the former to a huge loadstone; and in another work, not, however, published until many years after his death, he gives his opinions at greater length. In this treatise he asserts that the earth and the moon act upon each other like two magnets, but the influence of the earth must be greater than that of the moon, on account of its greater mass. Again, although the influence is magnetic in its nature it does not show itself as in ordinary magnets. "It is not," he says, "so as to make the bodies unite like two magnets, but that they may go on in a continuous course." In another place he ascribes the tides partly to the influence of the moon.

These speculations of the English philosopher had a strange fascination for Galileo. In 1608, he had the idea of recording their importance in a medal, which he proposed to strike on the occasion of the marriage of his pupil, Prince Cosimo, and the Archduchess Madeleine of Austria. Under the name *Cosmos,* he proposed to engrave the figure of the prince, and a magnet from which depended several pieces of iron, with the mottoes : *"Vim Facit Amor,"* and *"Magnus Magnes Cosmos."*

In the third "Day" of his *Dialogues on the Ptolemaic and Copernican Systems,* published in 1632, he utilises his own early magnetic observations, and warmly acknowledges the merits of Gilbert's work, declaring that his marvellous conception of the earth as a great loadstone was to him a subject of praise, admiration, and envy. The passage is worth quoting:

> I extremely praise, admire, and envy this author. I think him, moreover, worthy of the greatest praise for the many new and true observations that he has made, to the disgrace of so many vain and fabling authors, who write, not from their own knowledge, but repeat everything they hear from the foolish vulgar, without attempting to satisfy themselves of the same by experiment—perhaps that they may not diminish the size of their books.

Here he is evidently tilting at the "Scientific" writers, numerous in his and the following century, who filled their books with fables about the all-embracing powers and virtues of the magnet, and on which charlatans traded and grew fat. One such person he appears to have met in Venice, of whom he gives an amusing account in the same *Dialogues*, at the end of the first "Day." One of the speakers, Sagredo, commenting on the remarks of the previous speaker, says:

> You remind me of one who offered to sell me a secret art, by which, through the attraction of a certain magnetic needle, it would be possible to converse across a space of two or three thousand miles. And I said to him that I would willingly become the purchaser, provided only that I might first make a trial of the art, and that it would be sufficient for the purpose if I were to place myself in one corner of the room and he in the other. He replied, that in so short a distance the action would be scarcely discernible; whereupon I dismissed the fellow, saying that it was not convenient for me just then to travel into Egypt, or Muscovy,

> for the purpose of trying the experiment, but that if he chose to go there himself, I would remain in Venice and attend to the rest.

During his residence at Padua Galileo was in the habit of returning to Florence for the long summer holidays. On these occasions for some years, and beginning about 1601, he gave instructions in mathematics to the young prince, Cosimo (born 1590), there being, apparently, no one in Florence whom the Grand Duke thought capable of carrying on this branch of the boy's education. Galileo, of course, was duly sensible of the honour done in thus selecting him as the young prince's mathematical tutor; and the arrangement was desirable for other reasons—it added to his income, always a serious matter; and it gained him the esteem and friendship of the reigning family of Tuscany, which, although not able (as we shall see) to protect him entirely from misfortunes, was still strong enough to make his troubles lighter than they would otherwise have been.

While the Grand Duke was wont to declare that Galileo was the greatest mathematician in all Christendom, his wife, the Grand Duchess Christina, believed him to be the greatest of astrologers, and at the commencement of what proved to be her husband's last illness, she begged him to correct his horoscope! He did so, and communicated the result in a letter of 16th January 1609, according to which Ferdinando I had still many years to live. Galileo's prognostic was speedily proved to be false, as the Grand Duke died twenty-two days after!

Florence, from the earliest period, was noted for its cultivation of the rites of sacrifice and divination, and during the thirteenth and fourteenth centuries, astrology was taught in the Universities of Italy, and in particular at Padua and Bologna. Even in the succeeding three centuries, and in spite of advancing knowledge, judicial astrology still held sway over the hopes and fears, not only of the vulgar, but of the highest and best educated classes.

No wonder, then, if Galileo, *more temporum,* dabbled in horoscopes; but it is not to be supposed that a mind, which early discarded the trammels of ancient Sciences, and took nothing on trust or mere authority, could really have believed in them. We prefer to consider his action in this, and other instances, in the light of a pious fraud.

EARLY IN THE month of October 1608, the telescope was invented in Holland, and, according to all accounts, its discovery was the result of an accident. Long ago Epicurus defined the universe and all that it contains as the result of a fortuitous concourse of atoms, and so it is very often in the arts and Sciences of man, great discoveries are the result of the fortuitous juxtaposition of two and two. In the present case, and as the story goes, an apprentice, playing with spectacle lenses in the shop of one Hans Lipperhey, an optician of Middleburg, noticed that by holding two of them in a certain position a large and inverted view of objects was obtained. On hearing of this the master fixed two glasses in a tube, so that the weather-cock on a neighbouring church spire could be seen apparently nearer and upside down. This toy was shown in his window, where one day the Marquis Spinola chanced to see it, and entered the shop to examine it. Struck by the strange phenomenon it presented, he purchased the instrument, and afterwards gave it to Prince Maurice of Nassau, who thought it might be useful in military reconnaissances.

Leaving this part of the subject, the fact for us at present is that reports of the invention, more or less vague, spread slowly over Europe, and reached our philosopher while on a short visit to Venice about the middle of June 1609. As Galileo's independent invention of the telescope, like so many more of his discoveries, has been denied or belittled by envious detractors in his own and later times, the story had best be told in his own words. In a letter from Venice to his brother-in-law Landucci, dated 29th August 1609, he says:

> I write now because I have a piece of news for you, though whether you will be glad or sorry to hear it I cannot say, for I

have now no hope of returning to my own country, though the occurrence which has destroyed that hope has had results both useful and honourable. You must know then that about two months ago [ie. about June 1609] a report was spread here that in Flanders a spy-glass had been presented to Prince Maurice, so ingeniously constructed that it made the most distant objects appear quite near, so that a man could be seen quite plainly at a distance of 2 miles. This result seemed to me so extraordinary that it set me thinking, and as it appeared to me that it depended upon the laws of perspective, I reflected on the manner of constructing it, and was at length so entirely successful that I made a spy-glass which far surpasses the report of the Flanders one. As the news had reached Venice that I had made such an instrument, six days ago I was summoned before their Highnesses, the Signoria, and exhibited it to them, to the astonishment of the whole Senate. Many of the nobles and senators, although of a great age, mounted more than once to the top of the highest church tower in Venice, in order to see sails and shipping that were so far off that it was two hours before they were seen, without my spy-glass, steering full sail into the harbour; for the effect of my instrument is such that it makes an object 50 miles off appear as large as if it were only five.

Perceiving of what great utility such an instrument would prove in naval and military operations, and seeing that his Serenity the Doge desired to possess it, I resolved on the 24th inst. to go to the palace and present it as a free gift. On quitting the presence-chamber I was commanded to bide awhile in the hall of the Senate, whereunto the Procurator, Antonio Prioli, one of the heads of the University of Padua, came, and, taking me by the hand, said that the Senate, knowing the way in which I had served it for seventeen years at Padua, and being sensible of my courtesy in making it a present of the spy-glass, had ordered my election (with my good-will) to the Professorship for life, with a salary of 1,000 florins yearly; and as there remained yet a year to terminate the period of my last re-election, they willed that the

> increase of salary should date from that very day. Knowing that Fortune's wings are swift but that those of Hope are drooping (i.e. the hope of returning and settling in Florence), I said I was content to abide by his Serenity's pleasure. Then the illustrious Prioli, embracing me, said: 'As I command here and can order what I please (it being my turn this week), I will that after dinner the Senate assemble, and that your election be put to the ballot,' which was done [with few dissentient votes]. So I am bound here for life, and can only hope to enjoy a sight of my own country during the recesses.

In "Il Saggiatore," published in 1623, Galileo enters more fully into the reasonings which led him to the invention, and defends his right to consider the telescope as a child of his brain. He says:

> What part belongs to me in the invention of the telescope, and why may I reasonably call it my son? As I have long ago shown in my *Sidereus Nuncius,* news arrived at Venice, where I happened to be at the moment, that a Dutchman had presented to Count Maurice of Nassau a glass by means of which one could see distant things as clearly as if they were near. With this simple fact I returned to Padua, and, reflecting on the problem, I found the solution on the first night after my arrival, and the next day I made an instrument and reported the fact to my friends at Venice, with whom I had been discussing the rumour. In the next six days I made a more perfect instrument, with which I returned to Venice, and showed it for more than a month to the wonder and astonishment of the chiefs of the republic—a task which caused me no small fatigue. But perhaps it may be said that no great credit is due for the making of an instrument, or the solution of a problem, when one is told beforehand that the instrument exists, or that the problem is solvable. It may be said that the certitude of the existence of such a glass aided me, and

that without this knowledge I would never have succeeded. To this I reply, the help which the information gave me consisted in exciting my thoughts in a particular direction, and without that, it is possible they may never have been directed that way; but that such information made the act of invention easier to me I deny, and I say more—to find the solution of a definite problem requires a greater effort of genius than to resolve one not specified; for in the latter case hazard, chance, may play the greater part, while in the former all is the work of the reasoning and intelligent mind. Thus, we are certain that the Dutchman, the first inventor of the telescope, was a simple spectacle-maker, who, handling by chance different forms of glasses, looked, also by chance, through two of them, one convex and the other concave, held at different distances from the eye; saw and noted the unexpected result; and thus found the instrument. On the other hand, I, on the simple information of the effect obtained, discovered the same instrument, not by chance, but by the way of pure reasoning. Here are the steps: the artifice of the instrument depends either on one glass or on several. It cannot depend on one, for that must be either convex, or concave, or plain. The last form neither augments nor diminishes visible objects; the concave diminishes them, the convex increases them, but both show them blurred and indistinct. Passing then to the combination of two glasses, and knowing that glasses with plain surfaces change nothing, I concluded that the effect could not be produced by combining a plain glass with a convex or a concave one; I was thus left with the two other kinds of glasses, and after a few experiments I saw how the effect sought could be produced. Such was the march of my discovery, in which I was not assisted in any way by the knowledge that the conclusion at which I aimed was a verity.

But some people believe that the certainty of the result aimed at affords great help in attaining it. Let them read history, and they will find that Archites made a dove that could fly, and that Archimedes made a mirror that burned objects at great

> distances, and many other admirable machines. Now, by reasoning on these things such people, doubtless, will be able, with very little trouble and with great honour and advantage, to tell us how they were constructed. And even if they do not succeed, they will be able to certify for their own satisfaction that that ease of fabrication which they had promised themselves from the foreknowledge of the result is very much less than what they had imagined.

Of the first telescope referred to in the above extracts no further mention is made, so we may suppose that it was of little value; but the second, which he presented to the Doge and is unfortunately lost, is mentioned with some particulars in his *Sidereus Nuncius (Messenger of the Stars),* published at Venice about the middle of March 1610. It consisted of a leaden tube, with a plano-concave eye-glass and a plano-convex object glass, and had a magnifying power of 3 diameters, thus making objects appear three times nearer, and consequently nine times larger. From other sources we learn that the tube was about 70 centimetres long and about 45 millimetres diameter. This instrument was shown for the first time in public on 21st August 1609, from the top of the campanile of San Marco, when the farthest object that could be clearly seen was the campanile of the Church of San Giustina in Padua, distant about 35 kilometres in a straight line.

On his return to Padua Galileo made his third telescope, of which he only says that "it made objects appear more than sixty times larger," which is equivalent to a magnifying power of about 8 diameters. But in a very few days he had a much better one, which enlarged four hundred times. With this he made his first observations on the moon, which he "brought to a distance of less than 3 semi-diameters of the earth, thus making it appear about twenty times nearer and four hundred times larger than when seen by the unaided eye." He also turned it towards Jupiter, but with no specified results. To obviate the shaking when held in the hand, the instrument was firmly fixed on a support. The lenses were adjustable, the tubes which held them being capable of being drawn out of, or pushed into, the main tube. Thus, to see clearly things not very distant, the glasses should be drawn apart a little; while for very distant objects they had to be approached. He found he could grind large convex lenses more truly than small ones. He preferred therefore to make his object-glasses larger than necessary, and to cover a portion of their surface, leaving open around the centre just so large a space as he found, on testing, to give the best results.

While on a short visit to Florence, probably in October 1609, he had this instrument with him, and showed it to his late pupil, Cosimo II, now become Grand Duke, "who, to his great surprise and delight, was able to see that the moon was a body very similar to the earth."

Very early in January 1610, Galileo had constructed a fifth and still more powerful telescope, "sparing neither labour nor expense," which showed objects more than thirty times nearer and nearly one thousand times larger. With this instrument he not only verified and completed his observations on the moon, begun the previous autumn with his fourth telescope, but he also discovered Jupiter's moons, some of the fixed stars, and contributed to the solution of that long-standing puzzle to philosophers—the Milky Way.

Writing to Belisario Vinta, then with the Tuscan Court at Pisa, 30th January 1610, Galileo thus modestly alludes to his first series of discoveries:

> I am at present staying in Venice for the purpose of getting printed some observations which I have made on the celestial bodies by means of my spy-glass *(mio occhiale)* and which infinitely amaze me. Therefore do I give thanks to God, who has been pleased to make me the first observer of marvellous things unrevealed to bygone ages. I had already ascertained that the moon was a body very similar to the earth, and had shown our Serene Master, the Grand Duke, as much, but imperfectly, not then having such an excellent spy-glass as I now possess, which, besides showing the moon most clearly, has revealed to me a multitude of fixed stars never before seen, being more than ten

> times the number of those that can be seen by the unaided eye. Moreover, I have ascertained what has always been a matter of controversy among philosophers, namely, the nature of the Milky Way. But the greatest marvel of all is the discovery of four new planets. I have observed their motions proper to themselves and in relation to each other, and wherein they differ from the motions of the other planets. These new bodies move round another very great star, in the same way as Mercury and Venus, and, peradventure, the other known planets, move round the sun. As soon as my tract is printed, which I intend sending as an advertisement to all philosophers and mathematicians, I shall send a copy to his Highness, the Grand Duke, together with an excellent spy-glass, which will enable him to judge for himself of the truth of these novelties.

The tract referred to in the above letter is his *Sidereus Nuncius*, the preface of which is dated 4th March 1610, and the book, doubtless, appeared immediately after, say, towards the middle of March. In this epoch-marking treatise he gives the results of his observations to date, of which we proceed to give a *résumé*.

HIS OBSERVATIONS WERE FIRST DIRECTED to the moon. The discovery of new spots on its face, added to those already visible to the naked eye, and observations on the changes of light on those spots, led him to the conclusion that the surface of the moon, far from being smooth and polished, according to the opinion of the ancients, was rough with deep depressions and high mountains. Those parts which remained or became brilliant he inferred were land, like the solid parts of this earth, while those which remained obscure—the permanent spots—were water. The illuminated edges of the moon in all its phases showed themselves perfectly round, without those indentations which one would expect from the inequalities of its surface. Galileo explained this appearance—(1) by supposing that the mountainous parts, as it were, masked each other, so that at the distance of the earth the intervening depressions were not discernible, and (2) by the existence of a lunar atmosphere of a density such as to reflect the solar rays while not obstructing the vision. Thus the reflection of solar light by this atmosphere gave the appearance of a regular circular contour, only intensified in the parts most illuminated. From the appearance of illuminated mountain-tops in the dark part of the moon at some little distance from the broken line along which sunrise or sunset was general, he was able to judge of the height of some of these mountains. And his calculation agrees very well with the modern estimate. The higher mountains were found to rise 4 or 5 miles above the general level—a height which is seldom exceeded on the earth.

He, of course, remarked the feeble light, so-called phosphorescent, which, in the first and last quarters of the moon, makes visible to us that part of its disc which is no longer illuminated directly by the sun. After showing that this light did not originate in the moon itself, and was not reflected there from Venus, he concludes that it can only be due to the sunlight reflected from the earth to the moon, and thence reflected back to our eyes.

After referring for greater details to the work which he proposed to publish on the system of the world, he contends, contrary to the received opinion, that our earth is a moving planet, and that it exceeds the moon in luminosity, and, therefore, that it is far from being the sink of impurity hitherto supposed.

In examining the fixed stars and comparing them with the planets, Galileo discovered a remarkable difference. While the planets showed themselves as discs like little moons, the stars appeared no larger than they do to the naked eye, bright specks in the firmament, sending forth twinkling rays. In explanation of this fact he supposes that the telescope has the effect of stripping the star of the false light by which it is surrounded when viewed with the naked eye. This spurious corona is ascribed by him to the effect of irradiation which generally increases with the brightness of the field upon which the luminous object is projected. Thus, at sunset, when the obscurity of the heavens is tempered by the twilight, the stars, even of the first magnitude, appear very minute. So, with respect to Venus, notwithstanding her usual splendour, she does not exceed a star of the sixth magnitude on those occasions when she happens to be visible at noon.

Upon directing his telescope to the more conspicuous star-clusters, he was astonished to find that they contained, besides those already known, a great number of other stars too faint to be individually recognised by the naked eye. The number of the Pleiades, which had been fixed at seven, now rose to forty, while in the constellation of Orion, instead of seven he counted eighty stars. Certain portions of the Milky Way were resolved into a countless number of minute stars; and he inferred that as a whole it derived its singular whiteness from innumerable other stars which his instrument was not powerful enough to separate.

When Galileo turned his fourth telescope to the planets, he saw them as little moons. Jupiter's disc was of considerable magnitude, but in no other way did he differ from the other planets. Now, on 7th January 1610, directing his fifth and more powerful glass towards Jupiter, his attention was at once drawn to three small but very bright stars that appeared in his vicinity, two on the east side and one on the west. He at first imagined them to be fixed stars, and yet there was something in their appearance which he thought curious. They were all disposed in a right line parallel to the plane of the ecliptic, and were brighter than other stars of the same magnitude. This did not, however, induce him to alter his opinion that they were fixed stars, and therefore he did not note their distances from each other or from the planet. Happening, by mere accident, as he says, to examine Jupiter again on 8th January, he was surprised to find that the stars were now arranged quite differently. They were all on the west side, and were nearer to each other than on the previous evening, and at equal distances apart. The strange fact of the mutual approach of these stars had as yet no significance for him; it only excited his astonishment that the planet should be seen to the east of them all when on the previous night it was to the west of two of them. He very soon began to think that perhaps the motion of Jupiter might be direct, contrary to the accepted opinion of astronomers, and that he had got in advance of the stars. He therefore waited for the following night with some anxiety, but he was disappointed, for the heavens were enveloped in clouds. On 10th January he could see only two stars, and they were both on the east side! He suspected that the third might be concealed behind the disc of the planet. Those visible appeared as before in the same right line, and lay in the direction of the ecliptic. Unable to account for such changes by the motion of the planet, and being at the same time fully assured that he always observed the same stars, his doubts now turned into admiration, and he concluded that the motions must be referred to the stars themselves and not to the planet. He therefore determined to watch them with the closest attention.

On 11th January he again saw only two stars, still on the east side of Jupiter, but the outer one was now nearly twice as large as the other, although on the previous night they were almost equal. This fact, taken in connection with the constant change of the relative positions of the stars and the total disappearance of one of them, left no doubt on his mind of their real character. He concluded that there are in the heavens three stars revolving round Jupiter in the same way as Venus and Mercury revolve round the sun. On 12th January he again saw three stars, two on the east side of Jupiter, and one on the west. The third began to appear about three o'clock in the morning, emerging from the eastern limb of the planet; it was then very small, and discernible only with great difficulty. On 13th January he saw four stars, three of them on the west side and one on the east. They were all in a line parallel to the ecliptic, with the exception of the central one of the western group, which was a little towards the north. They were all about the same size, and shone with a much greater lustre than fixed stars of the same magnitude. January 14th was cloudy, but next night he saw all four stars to the west of the planet, all nearly in the same right line, and increasing in size and brilliancy, according to their distance from Jupiter.

And so he continued nightly, up to 2nd March 1610, to make these observations, sixty-six of which are figured and described in the *Sidereus Nuncius.*

The persistence of the relative distances between these four bodies and Jupiter in all their changes left no room for doubt that they accomplished with him, and in about twelve years, a revolution around the sun as a centre. Their own orbits round the planet were unequal in time, those nearest moving more rapidly than those more remote; while the most remote of all appeared to complete its revolution in one-half month.

"It is now," he says in conclusion, "not simply a case of one body (the moon) revolving around another body (the earth), while the two together make a revolution around the sun, as the Copernican doctrine teaches; but we have the case of four bodies or moons revolving round the planet Jupiter, as the moon does round the earth, while they all with Jupiter perform a grand revolution round the sun in a dozen years."

DURING THE EASTER recess in Padua, April 1610, Galileo, according to custom, would probably have visited Florence; but this time he had an express invitation from the Court, then at Pisa, to repair thither for the purpose of explaining to the Grand Duke his discovery of the four satellites of Jupiter, which, in honour of the reigning family of Tuscany, he proposed to call Medicean Stars, after the four brothers Cosimo II, Francesco, Carlo, and Lorenzo de Medici. Cosimo II, who all his life showed a sincere attachment to his old tutor, asked for and obtained the gift of the instrument with which this discovery was made; but Galileo quickly repented of his generosity. He evidently could not part with his "old discoverer," as he affectionately called it in after years; so, while always reserving it for the Grand Duke, he kept it near himself till his death, when it was handed over to Prince Leopoldo, brother of Ferdinando II.

5

GALILEO, PROFESSOR IN PADUA (CONCLUDED)

1592–1610

The hundred and more telescopes, which Galileo had made in the first half of 1610, were distributed with copies of the *Sidereus Nuncius* amongst the princes and learned men of Italy, France, Flanders, and Germany. The best instruments he reserved for particular friends and patrons, amongst whom he mentions the Duke of Bavaria, the Elector of Cologne, Cardinal del Monte, and the Duke of Urbino, as having "begged" for them. The Cardinal sent in return a small picture to which an indulgence was attached! The Duke of Bavaria was not behind-hand, but what his present was is not stated (let us hope it was more substantial than the Cardinal's); while the Elector of Cologne wrote that the pamphlet was disappointingly incomplete, since it contained no directions for the making of the instrument. He therefore requested Galileo to impart the secret, promising to recompense him in a princely fashion.

In communicating this request, and evidently not caring to bear the brunt of the Elector's anger in case of non-compliance, Galileo's brother, Michelangelo, wrote, 14th April 1610: "See if you can gratify the Elector by showing him how to make the instrument, and if not, write to him direct in your own way." Then he peevishly continues:

> You say not a word about the telescope *I* asked you for. If I am not a prince, able to remunerate you, at least I am your brother, and it seems very strange to me that you do not care to gratify me.

At the French Court the arrival of Galileo's telescope caused immense excitement, the queen, Marie de Medici, being particularly interested in it as the invention of a distinguished fellow-countryman. It is related that in her eagerness to see the moon through it, she would not wait till the instrument was suitably fixed at the open window, but fell on her knees on the floor, to the consternation of her suite and the amazement of the grave Italian in charge of the telescope.

The solicitude of the French Court to gain a place in the heavens by the side of the Medici family is very amusing. In a letter of 20th April 1610, the great astronomer is begged:

> In case you discover any other fine star, call it by the name of the Great Star of France, as well as the most brilliant of all the earth, and, if it seems fit to you, call it rather by his proper name, Henri, than by the family name Bourbon. Thus you will have an opportunity of doing a thing due and proper in itself, and, at the same time, of rendering yourself and your family rich and powerful for ever.

As the news of Galileo's marvellous discoveries spread over Italy, the popular excitement grew intense. Thus, in Florence, poets chanted the discoveries and the glory of their fellow-citizen, and a public fête was celebrated in his honour. In Venice, Girolamo Sirturo describes the excitement as amounting to frenzy, and tells an amusing story of his own experience. With the first telescope which he had succeeded in making, he ascended the tower of San Marco, in the hope of trying it unmolested. Unluckily for him, he was seen by some idlers in the square below, a crowd soon collected round him, who insisted on taking possession of the instrument, and, handing it one to another, detained him for hours till their curiosity was satisfied. Desirous of obtaining the same gratification for their friends, they endeavoured to find out where Sirturo lodged, but he, overhearing their enquiries, thought it better to quit Venice early the next morning and pursue his observations in a less inquisitive neighbourhood.

~~CHRIST FEEDING THE MULTITUDE~~

GALILEO FEEDING THE MULTITUDE WITH KNOWLEDGE

And they said unto him, "We have here but five hypotheses, and two theorems." He said, "Bring them hither to me."

IN THE *SIDEREUS NUNCIUS* GALILEO did not formally proclaim his discoveries in relation to, and in support of, the Copernican theory of the world; but in his lectures and conversations he made no secret of his belief. Nor, indeed, was any specific announcement needed; his readers could see for themselves the connection, and the speedy result was a tremendous explosion of incredulity and malice. The Aristotelians were furious, and even men like Welser of Augsburg, and Clavio of Rome, both admirers of Galileo, would not credit his statements until they had learnt better from observations of their own. The latter, who was the first mathematician of his day, for months, down to October 1610, "laughed at the idea of there being four new planets, to see which they must first be put inside the telescope. Let Galileo keep his opinions and welcome. I hold to mine." Nor did it mend matters when Galileo offered 10,000 scudi to any one who would construct so cunning an instrument. Others refused even to look through the telescope; some, lest they should see, others convinced they could not see, things of which Aristotle had made no mention!

Among other sticklers for conservatism were the celebrated professors, Cesare Cremonino of Padua, one of Galileo's colleagues, and Julius Libri of Pisa, both of whom peremptorily rejected, on a priori grounds, Galileo's discoveries and the conclusions he drew from them. Libri died in December 1610, refusing to look through a telescope, and stigmatising to the last the "absurdities" of the presumptuous Florentine. In communicating the news of Libri's death to his friend Welser (17th December), Galileo expressed the hope that this stiff-necked opponent of his "absurdities," who would not look at them from earth, might now perhaps see them on his way to heaven.

Some passages of Galileo's letter to Kepler of 19th August 1610 will best show how these men of Science refused to be convinced.

> You are the first and almost the only person, who, after a cursory investigation, has given entire credit to my statements. ... We will not trouble ourselves about the abuse of the multitude, for against Jupiter even giants, to say nothing of pygmies, fight in vain. Let Jupiter stand in the heavens and let the sycophants bark at him as they will. ... In Pisa, Florence, Bologna, Venice, and Padua, many have seen the planets, but all are silent on the subject and undecided; for the greater number recognise neither Jupiter nor Mars, and scarcely the moon, as a planet. ... What is to be done? Shall we side with Democritus or Heraclitus? I think, my Kepler, we will laugh at the extraordinary stupidity of the multitude. What do you say of the leading philosophers here to whom I have offered a thousand times of my own accord to show my studies, but who, with the lazy obstinacy of a serpent who has eaten his fill, have never consented to look at the planets, or moon, or telescope? Verily, just as serpents close their ears, so do men close their eyes to the light of truth. To such people philosophy is a kind of book, like the *Aeneid* or the *Odyssey*, where the truth is to be sought, not in the universe or in nature, but (I use their own words) by comparing texts! How you would laugh if you heard what things the first philosopher of the faculty at Pisa brought against me in the presence of the Grand Duke. He tried hard with logical arguments, as if with magical incantations, to tear down and argue the new planets out of heaven!

As a specimen of the "logical" arguments to which Galileo alludes in the above extract, this is what Francesco Sizzi, a Florentine astronomer, says in his *Dianoia Astronomica* (Venice, 1611.)

> There are seven windows given to animals in the domicile of the head, through which the air is admitted to the tabernacle of the

> body, to enlighten, to warm, and to nourish it. What are these parts of the microcosmos? Two nostrils, two eyes, two ears, and a mouth. So in the heavens, as in a macrocosmos, there are two favourable stars, two unpropitious, two luminaries, and Mercury undecided and indifferent. From this and many other similarities in nature, such as the seven metals, etc., which it were tedious to enumerate, we gather that the number of planets is necessarily seven. Moreover, these satellites of Jupiter are invisible to the naked eye, and therefore can exercise no influence on the earth, and therefore would be useless, and therefore do not exist. Besides, the Jews and other ancient nations, as well as modern Europeans, have adopted the division of the week into seven days, and have named them after the seven planets. Now, if we increase the number of the planets, this whole and beautiful system falls to the ground.

Another opponent deserves to be named, if only for the impudence of the charge he brings against Galileo.

"We are not to believe," says Christmann in his *Nodus Gordius,* "that nature has given Jupiter four satellites in order to immortalise the name of the Medici. These are the dreams of idle men who love ludicrous ideas better than our laborious maintenance of the heavens. Nature abhors such horrible chaos, and to the truly wise such vanity is detestable."

IN THE MIDST of all this opposition and abuse, we must not suppose that Galileo was without friends, and even some converts. Thus, on 7th May 1610, he was able to inform Secretary Vinta that even the most exalted persons in Padua, who had vehemently attacked him, had at length given up the game, and had acknowledged, *corain publico,* that they were not only convinced, but were ready to defend him against all comers.

The praises of Kepler, then renowned as the first astronomer in Europe, were, as we have just seen, a great consolation to him. Kepler had the *Sidereus Nuncius* at once reprinted in Prague, with a long and appreciative preface from himself, and some laudatory verses from Thomas Seggett, a learned Scotchman, a former pupil of Galileo, then working with him. In the preface, which is in the form of a letter to Galileo, dated 19th April 1610, Kepler says:

> I was sitting idle at home thinking of you, most excellent Galileo, and of your letters, when the news was brought me of the discovery of four planets by the help of the double eyeglass. ... The authority of Galileo had the greatest influence on me, earned by the accuracy of his judgment, and by the excellence of his understanding. So I immediately fell to thinking how there could be any addition to the number of the planets without upsetting my *Mysterium Cosmographicum,* published thirteen years ago, and according to which Euclid's five regular solids do not allow more than six planets round the sun. I am, however, so far from disbelieving the existence of the four circumjovial planets, that I long for a telescope to anticipate you, if possible, in discovering two round Mars (as the proportion seems to require), six or eight round Saturn, and, perhaps, one each round Venus and Mercury.

Galileo's detractors must have been hard pushed for a stick wherewith to strike him when they took Kepler's preface to be a covert attack. Certainly Maestlin, Kepler's old master, took it so, and wrote:

> In your essay you have plucked Galileo's feathers well—I mean you have shown him not to be the inventor of the telescope, not to be the first to observe the irregularities of the moon's surface,

> not to be the first discoverer of more worlds than the ancients were acquainted with, etc. One source of exultation was still left him, but from the apprehension of that Martin Horky has now entirely delivered me.

It is difficult to see where in Kepler's preface Maestlin found all this, for it is one continued encomium. Maestlin, however, was not the only one to misunderstand. The Martin Horky just mentioned, a young German travelling in Italy, wrote to Kepler: "I will never concede his four new planets to that Italian from Padua, though I die for it."

He followed up this declaration by publishing a book, which is evidently the one referred to by Maestlin. It professes to examine four main questions touching the alleged planets—(1) Whether they exist? (2) What are they? (3) What are they like? (4) Why they are? Horky summarily disposes of the first question by declaring that he had examined the heavens with Galileo's own glass, and could see no such thing as a satellite about Jupiter. As to the second he declares solemnly that he does not more surely know that he has a soul in his body, than that reflected rays are the sole cause of Galileo's observations. In regard to the third question he says, rather illogically, that these planets are like the smallest fly compared with an elephant; and as to the fourth he concludes that the only use of them is to gratify Galileo's vanity and thirst of gold. Horky sent a copy of this amazing production to Kepler, and, returning to Prague soon after, presented himself to the great astronomer. But the reception was little to his taste, for the burst of indignation which followed showed him Kepler's real sentiments. The conclusion is characteristic. After venting his wrath against "this scum of a fellow, whose obscurity had given him audacity," Kepler, recounting the story to Galileo, says:

> In the end, Horky begged so hard to be forgiven that I have taken him again into favour upon this one condition, to which he

has agreed—that I am to show him Jupiter's satellites, and that he is to see them, and own they are there.

AFTER COMPLETING his observations on Jupiter, Galileo turned his glass to the other planets to see if they, perchance, had attendant moons; but up to 25th June 1610, he was unable, with all his diligence, to discover any. At this he was inclined to be a little glad and a little sorry—glad because he would thus be the only one destined by God for so great a discovery; and sorry because he could not oblige the French Court by finding Henri Quatre a place in the heavens. However, before a month elapsed, he made another brilliant discovery —the ring of Saturn; only it did not appear to him as a ring, but like a triple star, of which the central one (Saturn itself) was three to four times larger than the laterals, and all three in a plane parallel to the equinoctial points or ecliptic. Not wishing to make public the discovery until he had made further observations in the autumn, when Saturn would be well above the horizon, and yet fearing that some one might forestall him, he announced the discovery in a brief letter, dated Padua, 30th July 1610, to Belisario Vinta at Florence, but begged him to keep it secret for a while. As a further precaution he sent to friends in Italy and Germany a jumble of thirty-seven letters as follows:

SMAISMRMILMEPOETALEUMIBVNENUGTTAVIRAS

Kepler and other friends puzzled long over this string of letters, the former thinking it had some reference to his favourite planet Mars. At length Giuliano de Medici, Tuscan ambassador at the German Court, was charged by the Emperor Rudolph II to ask for the solution, to whom Galileo, replying 13th November 1610, gave the following startling explanation:

Altissimum Planetam Tergeminum Observavi.

I have observed with great admiration that Saturn is not a single star but three together, which, as it were, touch each other. They have no relative motion, and are constituted in this form, the middle being much larger than the lateral ones. If we examine them with a glass of inferior power, the three stars do not appear very distinctly. Saturn has an oblong appearance somewhat like an olive, but by employing a glass which multiplies the *superficies* more than 1,000 times, the three globes will be seen very distinctly and almost touching, with only a small dark space between them.

I have already discovered a court for Jupiter, and now there are two attendants for this old man, who aid his steps and never leave his side.

The learned world of Italy had not yet had time to digest the surprising facts announced in the *Sidereus Nuncius,* when the asserted triple nature of Saturn again contravened the prevailing ideas that, by order of Aristotle, there was nothing new to be found in the heavens. Accordingly, the peripatetics were inclined to discredit the discovery; the most they would admit was that Saturn appeared to be of an oblong shape— precisely as Galileo said it appeared when viewed through glasses of less power than his own.

Continuing his observations, Galileo found that the lateral bodies did not retain the same apparent magnitudes. In fact, they had been gradually diminishing, although they appeared to be immovable, both with respect to each other and to the central body. They continued to grow less and less during the next two years, and towards the close of 1612 they vanished altogether! Horrified at this extraordinary phenomenon, and full of alarm for the consequences to himself when his Aristotelian opponents should come to hear of it, he thus wrote to Welser on 1st December 1612:

> Looking at Saturn within these last few days, I found it solitary without its accustomed stars, and, in short, perfectly round and defined like Jupiter, and such it still remains! Now what can be said of so strange a metamorphosis? Are, perhaps, the two smaller stars consumed like spots on the sun? Have they suddenly vanished and fled? Or has Saturn devoured his own children? Or was the appearance, indeed, fraud and illusion, with which the glasses have for so long mocked me and many others who have observed with me? Now, perhaps, the time is come to revive the withering hopes of those who, guided by more profound contemplation, have fathomed all the fallacies of the new observations, and recognised their impossibility. I cannot resolve what to say in a change so strange, so new, so unexpected. The shortness of time, the unexampled occurrence, the weakness of my intellect, the terror of being mistaken, have greatly confounded me.

However, he soon plucked up courage, and in the same letter conjectured that the two attendant stars would reappear after revolving round the planet, and that by the summer solstice of 1615, they would not only be again visible, but be more luminous, and larger.

Commenting on the above quoted passage, Arago and other modern writers after him conclude that Galileo was so discouraged by the disappearance of the lateral bodies that he made no further observations on Saturn. But Arago is most certainly wrong here. Galileo continued to observe Saturn for many years. By the middle of 1613, he was able to inform his friends that, according to his prediction, the lateral stars were reappearing!

Towards the end of his life, when Galileo was blind and a confirmed invalid from age and ailments, he once more turned his attention to Saturn. Castelli, writing to him from Rome under date 4th August 1640, said:

The other evening, on turning the telescope towards Saturn, I noticed, to my great amazement, that he was a single star, distinct and round, and with two other round stars (one at each side) lying in the direction from Levante to Ponente [i.e. nearly east and west], and no longer in the form of coifs attached to the central body, such as your first observations showed them.

Replying on 28th August 1640, Galileo wrote:

When first I observed Saturn he was composed of three round stars, situated in a straight line from Ponente to Levante, of which the central was much larger than the lateral ones. Thus I continued to see him for some months. Then, after an interval of some more months, I again examined him and found him solitary, i.e. the great central star was only to be seen. Amazed at this result, and supposing it to be due to some kind of change, I ventured to say that in five or six months, i.e. at the summer solstice, the two small lateral stars would reappear. They did, and so I saw them for a long time after. Then after another interval, during which Saturn was masked by the sun's rays, I again observed him, and now saw him with two mitres, instead of round stars, which gave him the figure of an olive. I saw the central globe very distinctly, and two very dark spots in the middle of the attachment of the mitres, or, as one may say, the ears. So I observed him for many years; and now your Rev. writes (as also other of my friends) that the mitres are transformed into two small globes. It may be that in the last three years, during which I have been unable to make any observations, Saturn may have become once again solitary, and then later on may have returned to the form in which I at first observed him. It will be for the future and for others to make observations, registering the times of mutation so as to accurately determine their periods

—that is, if there will be any persons curious enough to do what I, from the same motive (not knowing how to do better), have done for so long a time.

Thus we see that up to the last, Galileo had made no announcement as to the precise nature of Saturn's appendage. He contented himself with describing what he saw, and, recognising the incompleteness of his knowledge, and, perhaps, the inadequate power of his glasses, he left it to the future and to others to solve the problem.

6

GALILEO QUITS PADUA AND RETURNS TO FLORENCE

1610–1612

Galileo's fame, especially through his telescopic discoveries, and partly also through the exertions of his noisy opponents, had long extended beyond the bounds of Italy. The eyes of all Europe were directed to the great astronomer of Padua, and students flocked to him from all quarters. According to a familiar French proverb, one must suffer in order to be beautiful, so, to be a professor with a European reputation demands some sacrifice, entails some evil, which, under certain circumstances, may outweigh the good. It was so in Galileo's case; lectures and private lessons of all kinds left him little leisure for his own studies, and so, after twenty years' professorship at Pisa and Padua, he began to wish for a post in which he could devote himself entirely to the completion of various works on mechanics and astronomy, for which, during all these years, he had been amassing materials. A letter from Padua in the spring of 1609 shows his longing for this salaried leisure. It is not addressed, but from the context it must have been written to someone high in influence, if not in office, at the Court of Florence, probably to Belisario Vinta, the Grand Duke's chief Secretary of State. This first attempt, however, had no definite result, so that a few months later (after the invention of the telescope) he gratefully accepted, as we saw, the Chair of Mathematics at Padua for life. But the invention of the telescope and his consequent discoveries had now given him a world-wide reputation, and it appeared desirable to the Tuscan Court to attach to itself so great a man.

The first steps towards this end were taken when Galileo visited Pisa, about Easter 1610, in order to show to Cosimo II his telescopic discoveries, and especially the satellites of Jupiter, which bore his family name of Medici. Galileo's case is fully stated in his letter of 7th May 1610, to Vinta, as follows:

> I will not hesitate to say, having now laboured during twenty years, and those the best of my life, in dealing out (as one may say) in detail, and at the request of everybody, the little talent which God has given me, that my wish is to have sufficient leisure to enable me, before my life comes to a close, to conclude three great works which I have in hand, and which may, perhaps, bring some credit to me and to those who assist me in the undertaking, besides being a greater service to students than in the rest of my life I could do them by personal tuition. Greater leisure than I have here I doubt if I could get elsewhere, so long as I am obliged to give public and private lectures in order to meet the expenses of my family. Nevertheless, not even the liberty I have here is sufficient, when I am obliged to spend many and often the best hours of the day at the call of this and that man. My salary is 520 florins, which I am almost certain will be advanced to as many crowns upon my re-election, and this I can increase, by receiving pupils and by private lessons, to any extent I please. My public duties do not occupy me more than sixty half-hours in the year, and even then not so strictly but that I may, on occasion of any pressing matter, contrive to get some vacant days. The rest of the time is absolutely at my own disposal; but as my private lectures and domestic pupils take up very much of this spare time, to the hindrance of my own studies, I wish to be entirely exempt from public duties, and in a

great measure from the others. Therefore, if I am to return to my native country, I should wish that leisure and opportunity be afforded me to complete my works without employing myself in lecturing. In short, I wish to gain my bread by my writings, which I would always dedicate to my Serene Master.

Of useful and curious secrets I possess so many that their very abundance does me harm, for if I had but one I should esteem it greatly, and, perhaps, through it I might have found that fortune which as yet I have not met. And, indeed, I have not sought it. *Magna longeque admirabilia apud me habeo.* Many are no good to me, or I should say, they are only of use to princes; for they alone make wars, build fortresses, and, for their royal pleasure, spend such sums of money as private gentlemen cannot, any more than I can.

The works which I have to finish are chiefly—(1) two books on the system or structure of the Universe, an immense work, full of philosophy, astronomy, and geometry; (2) three books on local motion, a Science entirely new, no one, ancient or modern, having discovered any of the many admirable consequences which I demonstrate in natural and violent motions, so that I may with reason call it a new Science invented by me from its very first principles; (3) three books on mechanics, two on the demonstration of principles, and one of problems. Although others have treated this subject, no one either in quantity or quality has done a quarter of what I am writing on it. I have also treatises on natural and other subjects, such as (1) on sound and speech; (2) on light and colours; (3) on the tides; (4) on the composition of continuous quantity; (5) on the movements of animals; and others. I have also an idea of writing some books relating to the military art, giving not only a model of what a soldier ought to be, but teaching him with exact rules everything in the way of mathematics that it is his duty to know, as castrametation, manoeuvring battalions, fortifications, sieges, surveying, estimation of distances, knowledge of artillery, uses of various instruments, etc.

I also wish to reprint the 'Use of the Geometrical and Military Compass,' which is dedicated to his Highness, and is no longer to be procured. This instrument has met with such favour from the public that no others of the kind are now made, and I know that up to this several thousands of mine have been made and sold.

Then I need not say what an amount of labour will be required to fix the periods of the four new planets, a task the more laborious the more one thinks of it, as they are separated from one another by very brief intervals, and are all very similar in size and colour. So that, illustrious Signor, I must begin to think in what way I can free myself from those employments which so retard my own studies—particularly those, which another might fill quite as well as I can. Therefore, I beg you to bring these considerations to the notice of his Highness, and acquaint me with his decision.

As to salary, I shall be quite content with the sum you named to me at Pisa, feeling it an honour to be his Highness's servant. I say nothing as to the amount, being convinced that as I am to live upon it his Highness would not wish to deprive me of any of those comforts which others enjoy, who are less in want of them than I am. Therefore I say no more on this point.

Finally, as to the title and pretext by which I take service, I would wish that to the title of Mathematician his Highness would be pleased to add that of Philosopher, as I have studied a greater number of years in philosophy than months in pure mathematics. And how far I have profited by it, and if I can and ought to merit the title, I hope to be able to show his Highness as often as it is his pleasure to give me an opportunity of discussing such subjects with those whose knowledge is most esteemed.

This letter brought the business to a speedy settlement. On 5th June, Vinta wrote that Cosimo II was pleased to nominate him as First

Mathematician of the University of Pisa, with a yearly salary of 1,000 scudi, and without the obligation of residing at Pisa, or of delivering lectures. In reply, Galileo declared himself entirely satisfied with the proposed conditions, but added that he would like to be designated not only First Mathematician at Pisa, but also Philosopher and Mathematician to the Grand Duke himself. Accordingly, on 12th July 1610, the decree summoning him to Florence in this twofold capacity was issued.

Notwithstanding the many advantages which this new post secured to him, it was a bad exchange from the free soil of Republican Venice to the protection (ineffectual as it proved at the crisis of his life) of a princely house which, although personally well disposed towards him, could never shield him as could and would the Republic. About 1542, the Jesuits had established a school in Padua, and, increasing gradually in influence, had shown symptoms of a design to get the management of public education entirely into their own hands. After several violent disputes, it was at length decreed by the Venetian Senate, in 1591, that no Jesuit should be allowed to give instruction at Padua in any of the Sciences taught in the University. As years dragged on, the relations between Rome and Venice became increasingly strained, until at last, in April 1606, Pope Paul V took the extreme step of placing the contumacious Republic under an interdict. The Senate's reply was to expel forever the Jesuits from the soil of Venice. Since this event, full liberty of thought and teaching was enjoyed in the Republic; whereas, in religious Tuscany, Church influences were very strong and weighed heavily in all matters, and particularly in politics and Science.

EARLY IN SEPTEMBER 1610, Galileo left Padua, where, eighteen years before, he had found a ready welcome and an ever-increasing appreciation, deserting his staunch friends Paolo Sarpi, Francesco Sagredo, and many others. Indeed, he seems to have felt himself that he was not behaving well in this matter, for, in his letter to Vinta last quoted, he begged that the negotiations be kept secret until all was decided and therefore irrevocable.

Although he was received in Florence with much honour, soon the clouds of envy, malice, and bigotry began to form round him, and ultimately, as we shall see, combined, if not to his destruction, at least to embitter the rest of his life. Sagredo foresaw this clearly. On his return from the East in the spring of 1611, he wrote expressing his regret and disappointment at not finding Galileo, and his grave doubts as to the wisdom of the change.

"Where," he asks, "will you find the same liberty as in Venetian territory, and, notwithstanding all the good qualities of the young ruler of Tuscany, who can promise with any confidence that, if not ruined, you will not be persecuted and tossed on the surging billows of Court life by the raging storms of envy?"

Within less than a month after his arrival in Florence, Galileo made another astounding discovery in the heavens—namely, the varying crescent form of the planet Venus. He announced this discovery to his correspondent, Giuliano de Medici at Prague, in an anagram, as in the case of his observations on Saturn, as follows:

Haec immatura a me iam frustra leguntur O y.

After convincing himself, by three months' observations, of its correctness, he sent the following solution of the riddle on 1st January 1611:

Cynthiae figuras aemulatur mater amorum.

That is, Venus rivals the appearances of the moon; for Venus being now arrived at that part of her orbit in which she is between the earth and the sun, and with only a part of her enlightened surface turned towards us, the telescope shows her in a crescent form, like the moon in a similar position.

Tracing her through the visible portion of her orbit, Galileo had the satisfaction of seeing the illuminated part assume successively the forms appropriate to his hypothesis. It was with reason, therefore, that he laid stress on the importance of this observation, which established yet another fact obnoxious to the Aristotelians, namely, that a new point of resemblance was here found between the moon and one of the principal planets; for, as it was well known that the moon was luminous only when exposed to the sun's rays, so this change of figure in Venus demonstrated that she, and, therefore, probably all the other planets, were of themselves not luminous, but only reflected the sunlight which fell upon them. And thence he concluded that they must all revolve round the sun—"a fact which was surmised by Pythagoras, Copernicus, Kepler, and their disciples, but which could not be proved by ocular demonstration, as it now can be in the case of Venus and Mercury. Kepler and the other Copernicans may now be proud to have judged and philosophised correctly, and it may well excite disgust that they were regarded by the generality of men of book-learning as having little understanding and as not much better than fools."

It had always been a formidable objection to the Copernican theory that Venus and Mercury did not exhibit the same phases as the Moon, which they should do if they revolved round the sun. Copernicus himself had endeavoured to account for this, by supposing that the sun's rays passed freely through the body of the planets, and Galileo took occasion to praise him for not being deterred from adopting the system (which on the whole appeared to agree best with the phenomena) by meeting with some appearances which it did not enable him to explain.

Another objection, equally embarrassing to the Ptolemaic and Copernican systems, was raised. Why, it was asked, did not Venus appear four times as large when nearest to the earth as when farthest away? Galileo was now able to answer this also. Venus does not appear four times as large when she is nearest to the earth, simply because her illuminated part is not then four times as large, though her diameter is; and as with the naked eye we judge of her size only by the amount of light, the nearer distance is offset by the lesser light, so that her size does not seem to vary.

Of changes in Mars Galileo could say little for certain. After observing him carefully for four months he was seen to vary in size according to his distance from the sun. Galileo thought that when the planet was at the middle points of his orbit he observed changes in the illuminated disc something like the phases of the Moon and Venus; but it is difficult, he adds, with so small an object to say whether it is always perfectly round.

Of Mercury he also observed little, because that planet's orbit does not take him far from the sun, and, in consequence, his small disc is always so resplendent that not even the best telescope could deprive him of his factitious rays (irradiation).

IN CONSIDERATION of the intense interest, friendly and otherwise, excited by these epoch-making discoveries, and the probability of their being used against him in Rome, Galileo thought it desirable to go there himself, and acquaint at first hand the savants and dignitaries of the Church with his work in the heavens. These people, he argued, must be first made to see the facts with their own eyes, so that they may be able to comprehend and assent to the conclusions to be drawn from them. Therefore, on 15th January 1611, he wrote to Vinta (who was then with the Court at Pisa), informing him of his design, adducing the above reasons, and requesting the necessary leave of absence. His request was not only granted, but the Grand Duke placed a litter at his disposal, undertook to defray all expenses, and directed the Tuscan Ambassador in Rome to lodge him at the Embassy and to entertain him during the visit. Illness, however, supervened, so that it was not till 23rd March 1611 that he was able to set out, provided with many letters of introduction, amongst them one from Michelangelo the younger (nephew of the great sculptor and painter) to Cardinal Barberini (afterwards Pope Urban VIII).

He was received with the greatest distinction by princes and all the

Church dignitaries, who vied with learned laymen in doing him honour. Even those who discredited his discoveries and dreaded their results, were as eager as the true friends of Science to see and hear this wonder of the age. His first duty was to call on his old friends, Cardinal del Monte and the Jesuit Father Clavio, by the latter of whom he was presented to the leading Jesuits of the Roman College.

After exhibiting on several occasions in the garden of the Quirinal Palace, belonging to Cardinal Bandini, all his recent discoveries, or "celestial novelties" as they were called, a commission of four Scientific members of the Roman College was appointed, at the request of Cardinal Bellarmine, to examine and report. As the result, the commissioners, Fathers Clavio, Griemberger, Maelcote, and Lembo, were constrained to admit what they had long denied and ridiculed, being now convinced by ocular proofs of the truth of the facts announced by Galileo. By this opinion (given on 24th April) of the Papal experts, his discoveries received to a certain extent the sanction of the Church. Attentions of all sorts were now heaped upon him; Pope Paul V granted him a long audience, and assured him of his unalterable good-will, which, however, did not remain so, as the sequel will show. High dignitaries of the Church followed suit, and were lavish in their admiration; and the Accademia del Lincei (founded in 1603 by Prince Cesi), the prototype of our Royal Society, elected him a member. So that when he took his departure, on the 4th of June, Galileo left behind in Rome many sincere friends and admirers, and some very envious foes.

A letter of 31st May, from Cardinal del Monte to Cosimo II, will best show how Galileo's friends regarded his visit.

> Galileo has, during his stay in Rome, given great satisfaction, and I think he must have felt it no less himself, for he had the opportunity of showing his discoveries so well, that to all clever and learned men here they seemed no less true and well-founded than astonishing. Were we still living under the ancient republic

> of Rome I verily believe there would have been a column on the Capitol erected in his honour.

Immediately after the publication of the report of the Bellarmine Commission, Galileo announced a new discovery in the heavens, namely, dark spots on the body of the sun, which, towards the end of April 1611, he showed to several prelates and savants in Rome.

~~ENTRY OF JESUS INTO JERUSALEM~~
ENTRY OF GALILEO INTO ROME

And the multitudes that went before, and that followed, cried, saying, "Bravissimo to the Father of Modern Astronomy: Blessed is he that comes in the name of Science."

DESCRIBING these phenomena at a later date, Galileo states that at first he was undecided whether to explain the ever-changing nature and position of the spots by supposing that the sun revolved on his own axis, or by imagining that other unknown stars, besides Mercury and Venus, revolved about the sun, which were visible only as spots, and invisible at other times on account of their small distance from him. But further observation led him to abandon the latter supposition and to announce positively that the spots were in actual contact with the body of the sun, where they were continually appearing and disappearing, much as clouds about our earth; that their figures were very irregular, some being very dark and others less so; that one would often divide into two or three, and at other times two or three would unite into one; and that they all had a common and regular motion, being carried round with the sun, which turned on its axis in a little less than a lunar month, and in the same direction as the planets.

These observations were, in their consequences to Galileo, particularly unfortunate, as in the course of the controversy in which they entangled him, he first became personally embroiled with the powerful Jesuit order, whose influence was one of the chief causes of his subsequent misfortunes.

A Jesuit father, Christopher Scheiner, Professor of Mathematics at Ingolstadt, claimed priority in the discovery of the sun-spots, asserting that early in 1611 he first noticed them and showed them to his pupils. Scheiner stated his case in three letters over the signature *Apelles latens post tabulam,* and addressed to Mark Welser, Chief Magistrate of Augsburg. In the first letter, dated 12th November 1611, he states that he made his first observations seven months previously; but apparently he then attached no value to them, for, on resuming his work in the following October, he thought the appearances arose from some defect in his glasses—an idea which was only dissipated by these further studies. Clearly then, it was only *after* resuming his observations (October 1611) that he believed in the actual existence of the spots, which he then explained by supposing them to be caused by multitudes of little planets, revolving round the sun in an orbit inside Mercury, and producing the appearance of spots in crossing his disc.

On the publication of Scheiner's letters at Augsburg, Welser, on 6th January 1612, sent a copy to Galileo requesting to be favoured with his opinions of the phenomena therein described. Galileo replied in three letters in which he easily combated Scheiner on the nature of the spots, and stated that his own observations dated back eighteen months.

In these letters Galileo further stated that the spots often dispersed like vapours or clouds, which he supposed them to be; that they sometimes lasted only one or two days, and at other times thirty or forty days; that they contracted in breadth as they approached the sun's limb, without diminishing their length; that they described circles parallel to each other; that the rotation of the sun again brought the same spots into view; and that they were never seen to extend to a greater distance than 30° from the sun's equator.

Besides the spots, Galileo's telescope disclosed other interesting appearances on the sun's disc. Some parts are perceived to be brighter than the rest of the surface, and hence are called *faculae*. These phenomena are described in the third letter to Welser, where with great acuteness he adduces them as a proof that the spots are attached to the surface of the sun, and are not little planets as Scheiner supposed; for the telescope frequently showed spots and faculae travelling across the sun's disc together; and, accepting Scheiner's supposition with regard to the former, it would be probable that the latter also should sometimes appear as bright spots beyond the sun's limb—a conclusion which was totally at variance with all his observations.

These letters were written from the Villa delle Selve of his friend Filippo Salviati, near Signa (9 miles from Florence), where Galileo spent much of his time at this period, and particularly during his frequent indispositions, as he considered that the air of the city was prejudicial to his complaints.

The letters, with Scheiner's appended, were printed at Rome in

March 1613, and were dedicated to Salviati. They were published under the auspices of the Accademia dei Lincei, "in order," as the Academy report sets forth, "to mark its sense of the merit of the work and the claim of its author to be regarded as the first discoverer of Sun-spots."

IN ORDER TO complete our history of the sun-spots in so far as Galileo is concerned, we have left unnoticed other important matters on which he was at the same time engaged, and to these we must now return. Soon after his return to Florence in June 1611, he wrote a pamphlet in the usual form of a letter, dated 1st September 1611, and addressed to Father Griemberger, "on the Inequalities of the Moon's surface," or lunar mountains. The moon was with him a stock subject for observation, the results of which he communicated from time to time in letters to (besides Griemberger) Welser, Brengger, Gallanzoni, and others. Indeed, his last astronomical discovery, towards the close of life and just before he became blind, was connected with the moon.

It had been asserted that, as the full moon always presented a well-defined outline, whether when viewed with the naked eye, or through a telescope, it was impossible that there could exist any equalities around her circumference. Galileo, however, maintained that the irradiation of the moon's light, by obliterating any asperities around her edge, might effectually conceal the real nature of that part of her surface. With respect to irradiation generally, he remarked that it increases with the brightness of the object. It is from this cause that the planets near the sun have a greater irradiation than those that are more remote. So intense is the irradiation of Mercury, that it is impossible, even with the most powerful telescope, to deprive him of his brilliant corona. The same is true, though in a less degree, with respect to Mars. On the other hand, Jupiter, and especially Saturn, being more feebly illuminated by the solar light, lose their irradiation in the telescope, and disclose their true figures. With respect to Venus, when she is near her inferior conjunction, she resembles the new moon; but such is the intensity of her irradiation, that she appears to the naked eye like any other star. In this position, however, as the extent of the illuminated surface is small, and the light is at the same time enfeebled by the obliquity of the surface, it is possible by means of a telescope to discern the real appearance of the planet. When, however, she is near her superior conjunction, she presents a complete hemisphere of vivid light towards the earth of such intensity that even the most perfect telescope does not suffice to destroy her irradiation and reveal to us her true figure.

Galileo, therefore, contends that since the effect of irradiation is so great as to conceal from the unaided eye the immense cavity of Venus when she assumes the form of a crescent, it is much more probable that a telescope will fail to efface the irradiation of the moon enough to disclose the small eminences and cavities which may be situated near the edge of her disc.

Like all subsequent observers up to the close of the eighteenth century, Galileo is said to have believed in the existence of lunar seas and of a lunar atmosphere; while on the question of the existence of lunar inhabitants he is supposed to have kept an open mind. His belief in a lunar atmosphere may be conceded, but as regards the other points the very reverse is the fact, as the following letter shows. Writing to Giacomo Muti from Rome, on 28th February 1616, Galileo says:

> A few days ago, when paying my respects to the illustrious Cardinal Muti, a discussion arose on the inequalities of the moon's surface. Signor Alessandro Capoano, in order to disprove the fact, argued that if the lunar *superfecies* be unequal and mountainous, one may say as a consequence that, since Nature has made our earth mountainous for the benefit of plants and animals beneficial to man, so on the moon there must be other plants and other animals beneficial to other intellectual creatures. Such a consequence, he said, being most false, therefore the fact from which it is drawn must also be false, therefore lunar mountains do not exist! To this I replied: As to

the inequalities of the moon's surface we have only to look through a telescope to be convinced of their existence; as to the 'consequences,' I said, they are not only not necessary, but absolutely false and impossible, for I was in a position to prove that neither men, nor animals, nor plants as on this earth, nor anything at all like them can exist on the moon. I said then, and I say now, that I do not believe that the body of the moon is composed of earth and water, and wanting these two elements we must necessarily conclude that it wants all the other things which without these elements cannot exist or subsist. I added further: even allowing that the matter of the moon may be like that of the earth (a most improbable supposition), still not one of those things which the earth produces can exist on the moon, since to their production other things besides earth and water are necessary—namely, the sun—the greatest agent in Nature—and the resulting vicissitudes of heat and cold, and of day and night. Now, such vicissitudes are on the moon very different from those on the earth. In the latter case, to produce a diversity of seasons, the sun rises and falls more than 47° (in passing from one tropic to the other); in the former case the variation is only 5° on each side of the ecliptic. While, therefore, on the earth the sun in every 24 hours illuminates all parts of its surface, each half of the moon is alternately in sunshine and darkness for 15 continuous days of 24 hours. Now, if our plants and animals were exposed to ardent sunshine every month for 360 consecutive hours, and then for a similar time were plunged in cold and darkness, they could not possibly preserve themselves, much less produce and multiply. We must, therefore, conclude that what would be impossible on our earth under the circumstances we have *supposed* to exist, must be impossible on the moon where those conditions *do* exist.

NO SOONER WERE these questions of sun-spots and lunar mountains out of hand than Galileo became engaged in another protracted discussion, which resulted in another famous treatise, published in Florence in 1612. This is his *Discourse on Floating Bodies,* in which he uses the principle of virtual velocities from which modern geometers, and especially Lagrange, have drawn largely, and by the aid of which he demonstrates the more important theorems of hydrostatics.

The question of floating bodies had been discussed at one of the Scientific parties which the Grand Duke liked to assemble round him, and the general opinion appearing to be that of Aristotle, that the floating or sinking of a body depended principally upon its shape, Galileo undertook to show its untenableness. Someone in the company asserted that condensation is the effect of cold, and ice was mentioned as an instance. Galileo retorted that ice is rather water rarefied than water condensed, the proof of which is that it always floats upon water. His opponents rejoined that this phenomenon was due, not to the lightness of the ice, but to its incapacity, owing to its flat shape, to overcome the resistance which the water opposed to its sinking. Galileo denied this, and asserted that ice of any shape would float upon water, and that if a flat piece of ice were forced to the bottom it would, when left to itself, rise again to the surface.

The general character of this remarkable Discourse will be understood from the following passages:

> The diversity of figure given to any solid cannot be the cause of its floating or sinking in water; so that if a solid made, for instance, into a sphere sinks or floats, it will likewise sink or float when made into any other form. The breadth of the figure may indeed retard its velocity as well of ascent as of descent, and more and more in proportion to the breadth and thinness of the said figure; but that it can be reduced to such a form as to put an end to its motion I hold to be impossible. In this I have met with opponents who, pointing to some experiments, and, in particular, taking a thin board of ebony and a ball of the same wood, and

> showing that the latter sinks, while the former, if placed lightly on the surface, floats, hold by the opinion of Aristotle that the cause of flotation is the breadth of the figure, which renders it unable to overcome the resistance of the water to penetration; whereas in the case of the ball its form enables it to overcome this resistance, and it readily sinks. I assert, on the contrary, that there is not any solid of such lightness or of such a figure that, being placed on the water, will not divide and penetrate it; and if you examine carefully your thin boards of wood you will see that they have part of their thickness under water; and, moreover, you will see that shavings of ebony, stone, and metal, when they float, have not only broken the continuity of the water, but are with all their thickness under the surface, and this more and more according to their specific gravity; so that a thin plate of lead will be lower than the surface of the surrounding water by at least twelve times the thickness of the plate, and a gold plate will dive below the level almost twenty times its thickness.

In order to show more clearly the non-resistance of water to penetration, Galileo then directs a cone to be made of wood or wax, and asserts that when it floats, either with its base or its apex in the water, the solid content of the part immersed will be the same, although the apex is (by reason of its shape) better adapted to overcome the resistance of the water to division. The shape, then, cannot be the cause of the buoyancy. Or the experiment may be varied by mixing the wax with lead filings till the compound sinks in water, when it will be found that in any shape the same weight of cork must be added before it will rise to the surface. He goes on:

> This silences not my antagonists, who say that my experiments import little to them, and that it is enough for them to have demonstrated in one instance, and in such manner as pleases them best, that an ebony ball when put into water sinks to the

bottom, while an ebony board stays to swim at the top, and the matter being the same, and the two bodies differing in nothing but in figure, they affirm that with all perspicuity they have demonstrated and sensibly manifested what they undertook. Nevertheless, I believe and think I can prove that this very experiment proves nothing against my theory. And first, it is false that the ball sinks and that the board does not; for the board will sink too, if you do to both as the words of our question require; that is, if you put them both in the water; for to be in the water implies to be placed in the water; and, by Aristotle's own definition of place, to be placed imports to be environed by the surface of the ambient body. But when my antagonists show the floating board of ebony they put it not into the water, but upon the water, where, being detained by a certain impediment (of which more *anon*), it is surrounded partly by water, partly by air, which is contrary to our agreement, for that was that the bodies should be in the water, and not part in the water, part in the air. I will not omit another reason, founded upon experience, and, if I deceive not myself, conclusive against the notion that figure and the resistance of the water to penetration have anything to do with the buoyancy of bodies. Choose a piece of wood or other matter, as for instance walnut-wood, a ball of which rises from the bottom of the water to the surface more slowly than a ball of ebony of the same size sinks, so that clearly the ball of ebony divides the water more readily in sinking than does the walnut in rising. Then, take a board of walnut-wood equal to and like the floating ebony of one of my antagonists, and if it be true that this latter floats by reason of the figure being unable to penetrate the water, the other of walnut-wood, without all question, if thrust to the bottom ought to stay there, as having the same impeding figure, and in consequence unable to overcome the said resistance of the water. But if we find by experience that not only the thin board, but every other figure of the same walnut-wood, will return to float, as unquestionably we shall, then I must desire my opponents to forbear to attribute the floating of the ebony to the figure of the board, since the resistance of the water is the same in rising as in sinking, and the force of ascension of the walnut-wood is less than the ebony's force for going to the bottom.

Now, let us return to the thin plate of gold or silver, or the thin board of ebony, and let us lay it lightly upon the water, so that it may stay there without sinking, and carefully observe the effect. It will appear clearly that the board or plate is lower than the surface of the water, which rises up and makes a kind of rampart round it on every side. But if it have already penetrated and overcome the continuity of the water, and is of its own nature heavier than the water, why does it not continue to sink, but stop and suspend itself in that little dimple that its weight has made in the water? My answer is, because in sinking till its surface is below the water, which rises up in a bank round it, it draws after and carries along with it the air above it, so that that which descends is not only the board of ebony (or metal plate), but a compound of ebony and air, from which composition results a body no longer specifically heavier than the water, as was the ebony or metal alone. But, gentlemen, we want the same matter; you are to alter nothing but the shape, and therefore have the goodness to remove this air. This may be done simply by washing the upper surface of the board, for the water having once got between the board and the air will run together, and the ebony will sink to the bottom; and if it does not, you have won the day. But methinks I hear some of my antagonists cunningly opposing this, and telling me that they will not on any account allow their board to be wetted, because the weight of the water so added, by making it heavier than it was before, forces it to the bottom, and that the addition of new weight is contrary to our agreement.

To this I answer first, that nobody can suppose bodies to be put into water without wetting them, nor do I wish to do more to the board than you do to the ball. Moreover, it is not true that the board sinks on account of the weight of the water added in the washing; for I will put ten or twenty drops on the floating board, and so long as they stand separate it shall not sink; but if the board be taken out, and all that water wiped off, and the whole surface be bathed with a single drop, and be again put upon the water, there is no question but it will sink, water running to cover it, being no longer hindered by the air. In the next place, it is altogether false that water can in any way increase the weight of bodies immersed in it, for water has no weight in water, since it does not sink. Now, just as he who should say that brass by its own nature sinks, but that when formed into a kettle it acquires from that figure a virtue of lying in the water without sinking, would say what is false, because that is not purely brass which then is put into the water, but a compound of brass and air; so is it neither more nor less false that a thin plate of brass or ebony swims by virtue of its dilated and broad figure.

Some may wonder that I affirm this power to be in the air of keeping a board or metal plate from sinking, as if in a certain sense I would attribute to the air a kind of magnetic virtue for sustaining heavy bodies with which it is in contact. To satisfy all these doubts, I have contrived the following experiment to demonstrate how truly the air does support these solids; for I have found, when one of these bodies (which floats when placed lightly on the water) is thoroughly bathed and sunk to the bottom, that, by carrying down to it a little air, without otherwise touching it in the least, I am able to raise and carry it back to the top, where it floats as before. To this effect I take a ball of wax, and with a little lead make it just heavy enough to sink very slowly to the bottom, taking care that its surface be quite smooth and even. This, if put gently into the water, submerges almost entirely, there remaining outside only a very little of the top, and, so long as it is thus joined to the air, the ball floats; but if we take away the air, by wetting this top, the ball sinks to the bottom, and remains there. Now to make it return to the surface by virtue of the air which before sustained it, thrust

> into the water a glass, with the mouth downwards, which will, carry with it the air it contains; move this downwards towards the ball, until you see that the air has reached the top of it; then gently draw the glass upwards and you will see the ball rise, and afterwards stay on the top of the water, if you carefully withdraw the glass without too much disturbing the water. There is therefore a certain affinity between the air and other bodies, which holds them united, so that they separate not without a kind of violence, just as between water and other bodies; for in drawing such bodies wholly out of the water, we see it follow them, and rise sensibly above the level before it quits them.

His opponents were in hopeless confusion between the phenomena of hydrostatic pressure, which they professed to be discussing, and those of capillary action which they quoted against Galileo. With his careful observations to support him, Galileo would have carried conviction more readily had he emphasised the distinction between the two questions, and realised himself that the floating plate of metal indicated a natural property of liquids which deserved special investigation.

Having established his theory of buoyancy by these ingenious experiments, Galileo proceeds to show what must be the dimensions of a plate of any substance which will float as the wax does, assuming in each case that we know the greatest height at which the rampart of water will stand round it. In like manner he shows that a pyramidal or conical figure may be made of any substance in such a way that, by the help of the air, it shall rest upon the water without wetting more than its base; and, furthermore, that we may form a cone of any substance so that it shall float (partially submerged) if placed gently on the surface with its point downwards, whereas no care or pains will enable it to do so with its base downwards, owing to the different proportions of air which in the two positions are in contact with the water.

As may be gathered from the foregoing excerpts, the book contains many ingenious experiments and much acute reasoning in support of the true principles of hydrostatics. Like all his other works it encountered violent opposition, and Galileo had, more than once, to enter the field to repel the counter "arguments" of his opponents. The first published attack was by Giorgio Coresio, a professor in Pisa, under the form of a letter to the Archbishop of Florence. The reply to this was entrusted to Galileo's valiant disciple and champion, Benedetto Castelli. The manuscript was submitted to the Master and revised by him, and was about to be published when Coresio was dismissed (on its leaking out that though professedly a Catholic he really belonged to the Greek Church), and died soon after. Thereupon, the work was laid aside, on the principle *de mortuis nil nisi bonum.* A second attack was published anonymously, but was well known to be the work of Arturo d'Elci, head of the University of Pisa, the same who, as we shall presently see, forbade any allusions in lectures to the astronomical discoveries of Galileo. To this Galileo replied in a long letter, dated 4th January 1613, and addressed to Tolomeo Nozzolini. More elaborate attacks were next published by Lodovico delle Colombe and Vincenzio di Grazla. To these a detailed and overwhelming answer was printed in the name of Benedetto Castelli. After pulverising all the "arguments" of his opponents, the writer tauntingly bids them remember that he was merely Galileo's pupil, and thence to conclude how much more effectually Galileo himself would have confuted them had he thought it worth his while.

It was in reference to this controversy that Galileo declared that ignorance had been the best master he ever had, since, in order to be able to demonstrate to his adversaries the truth of his conclusions, he had been forced to prove them by such a variety of experiments as made himself doubly confident; though to satisfy his own mind alone he had never felt it necessary to make many.

7
FIRST APPEARANCE BEFORE THE INQUISITION

1612–1617

The uncompromising boldness with which Galileo published and supported his opinions had, as we have seen so often, raised against him a host of enemies, who each had objections peculiar to himself, but who now began to perceive the policy of uniting their strength in the common cause to crush, if possible, so strange an innovator. The Aristotelian professors, the Jesuits, the political churchmen, and those timid but respectable persons who at all times dread innovation, whether it be in religion or in Science, all were drawn together against the philosophical tyrant who threatened them with the penalties of too much knowledge.

The party of Galileo, though weak in numbers, was not without power and influence. He had trained around him a devoted band who idolised his genius, and disseminated his views, for many of his favourite pupils were now professors in the leading schools of Italy.

No longer able to combat Galileo's hard facts and powerful arguments by asserting that the facts were faults in his glasses and instruments devilishly designed, and that his arguments were vain-glorious and "Scientifically" absurd, his opponents now took their stand on theology, and raised the cry of the Church in danger. On the simple ground that his astronomical doctrines were incompatible with

Scripture, a great din was raised and began to make itself heard, especially in Florence, early in the year 1612.

With the first mutterings of the storm in his ears, Galileo wrote thus to Prince Cesi on 12th May in that year:

> I suspect that this new discovery [the sun-spots] will be the signal for the funeral, or rather for the last judgment of the pseudo-philosophy—the funereal signals having already been shown in the moon, the Medicean stars, Saturn, and Venus. And I expect now to see the peripatetics put forth some grand effort to maintain the immutability of the heavens!

True to her original attitude of hostility, Galileo's *Alma Mater,* Pisa, was consistent and persistent in her opposition to his teachings. Thus Father Castelli, who in October 1613 was called to the Mathematical Chair in that University, told Galileo (6th November 1613) that he was forbidden to treat in his lectures of the double motion of the earth, or even to hint at its probability! "Of our controversies," he goes on to say, "not a word whatever, a thing which astonishes me. Your marvellous discoveries are scarcely known here even by name!"

Notwithstanding all this commotion in the schools and amongst the minor churchmen, the authorities at Rome had not yet taken the alarm. On the contrary, we find the Cardinals Maffeo Barberini (afterwards Pope Urban VIII) and Federigo Borromeo thanking Galileo for sending them his book on the sun-spots, and expressing their admiration for his researches as described therein. Similarly, Battista Agucchia, one of the principal officials of the Papal Court, writing 8th June 1613, expressed his belief that Galileo's teachings would in time be universally acknowledged, "although now they had many opponents, partly from their novelty and extraordinary character, and partly from the envy and obstinacy of those who had from the first maintained the contrary opinion."

But amongst his other friends (not in Rome) there was at least one more discriminating, one who knew better the men and their spirit—Fra Paolo Sarpi of Venice. This old and faithful friend did not share this confidence in the triumph of truth—"I foresee," said he, "that the ecclesiastical authorities will change a question of physics and astronomy into one of theology, and that to my great grief Galileo, if he wants to live in peace and escape the charge of heresy, will have to recant." "The day will come," he continues, "when men of Science will have to deplore the disgrace of Galileo." Prophetic words, as the sequel will show.

AN ACCIDENTAL CIRCUMSTANCE was the spark which fired the train. One day in December 1613, Castelli and other learned men were guests at the Grand Duke's table at Pisa, where, as usual, the Tuscan Court was wintering. The conversation turning on the "satellites of Jupiter," Castelli took the opportunity of extolling and expatiating on his master's discoveries. One of the guests, Boscaglia, Professor of Physics at the University, and a peripatetic of the purest water, managed to excite the religious scruples of Christina, Dowager Grand Duchess, by telling her that all Galileo's telescopic discoveries were true, only the deduction from them of the double motion of the earth must be wrong, as the Holy Scriptures were clearly opposed to such a doctrine. Castelli, who had left the apartment, was recalled to answer this objection. At first he deprecated bringing the Bible into the controversy, but as this was unavailing, he resolutely took the theological standpoint, and defended the new views of the universe so well, that many of those present took his side—the Dowager Duchess standing alone, and Boscaglia taking no part. Castelli hastened to apprise Galileo of this incident, and added that it appeared to him that the Grand Duchess had merely persisted in opposition, in order to draw him out (Letter of 14th December 1613).

This, then, was the provocation to Galileo's famous letter of 21st December 1613 to Castelli, in which for the first time he engages in theological discussions, and which, although not intended for publication, was to be turned to account by his enemies, and to form the groundwork of his subsequent trial. The letter is too long to quote in full, but the following passages will show the line of argument pursued. He begins by showing that there is as much difficulty in reconciling the Ptolemaic as the Copernican system of the world with the astronomical expressions contained in the Scriptures, and asserts that—the object of Scripture not being to teach astronomy—such expressions are there used as would be intelligible to the vulgar, and without regard to the true structure of the universe. He then goes on:

> As the Bible, although dictated by the Holy Spirit, admits (for the reasons given above) in many passages of an interpretation other than the literal one, and as, moreover, we cannot maintain with certainty that all interpreters are inspired by God, I think it would be the part of wisdom not to allow any one to apply passages of Scripture in such a way as to force them to support as true any conclusions concerning nature, the contrary of which may afterwards be revealed by the evidence of our senses, or by actual demonstration. Who will set bounds to man's understanding? Who can assure us that everything that can be known in the world is known already? ... I am inclined to think that Holy Scripture is intended to convince men of those truths which are necessary for their salvation, and which being far above man's understanding cannot be made credible by any learning, or by any other means than revelation. But that the same God who has endowed us with senses, reason, and understanding, does not permit us to use them, and desires to acquaint us in another way with such knowledge as we are in a position to acquire for ourselves by means of those faculties— that, it seems to me I am not bound to believe, especially concerning those Sciences about which the Holy Scriptures contain only small fragments and varying explanations; and this is precisely the case with astronomy, of which there is so little that the planets are not all enumerated, only the sun and moon, and once or twice Venus

> under the name of Lucifer. This, therefore, being granted, I think that in discussing natural phenomena we ought not to begin with texts from Scripture, but with experiment and demonstration, for from the Divine Word Scripture and Nature do alike proceed. And I can see that that which experience sets before our eyes concerning natural effects, or which demonstration proves unto us, ought not upon any account to be called in question, much less condemned, upon the testimony of Scriptural texts, which may (under their mere words) have meanings of a contrary nature. ...

The letter concludes with a long discussion of Joshua's miracle, ending in a *reductio ad absurdum*.

Castelli, seeing nothing objectionable in this letter, gave it a wide circulation in written copies. But not so the enemies of Galileo, who eagerly seized it as a weapon to be used against him. They turned his emphatic opinion that the Scriptures had no business in Scientific controversy into the reproach that he assailed the universal authority of the Bible, and, referring to his explanation of Joshua's miracle, they loudly demanded that the Scriptures must be protected from the arbitrary interpretations of profane laymen.

After some months of underhand agitation, Father Caccini of the Dominican convent of San Marco was the first to declare war openly, in a sermon from the pulpit of Santa Maria Novella in Florence. Preaching on the fourth Sunday in Advent (December) 1614, and selecting as his texts chap. x. vers. 12, 13, of Joshua, and chap. i. ver. 11, of Acts, he began with the words: "Ye men of Galilee! why stand ye gazing up into heaven?" He asserted that Galileo's doctrine of the earth's revolution round the sun was irreconcilable with the holy Catholic faith, since it contradicted several passages in Scripture. He reminded his hearers that no one was permitted to interpret the Bible in any other sense than that adopted by the Fathers; he as good as denounced Galileo's teachings as heretical; and wound up with a coarse attack on mathematicians in general, whose Science he called an invention of the devil, and who should be banished from all Christian states, since all heresies proceeded from them. A preacher at the Duomo (cathedral), a Jesuit strange to say, replied, and undertook to show that Copernicus was right; that Galileo and his followers were good Catholics; and that Caccini and the Dominicans generally were ignorant fools! As may be imagined, this line of argument added fuel to the fire, and the result was a great commotion in Florence. This fierce attack made Galileo lose patience; he wished to publicly resent the insults, but on the advice of friends he kept quiet, and the storm subsided for a while.

THE MAN who first brought Galileo's affairs before the Inquisition (or the Holy Office) was one Father Lorini, a friend of Caccini and a member of the same convent. This man was Professor of Ecclesiastical History in Florence, and was from the beginning of 1612 a ringleader of all the local intrigues against Galileo. A copy of the letter to Castelli had come into his hands, and this, after Caccini's fierce sermon, he sent to the Holy Office in Rome with an unsigned denunciation, dated 5th February 1615, and addressed to Cardinal Mellini, President of the Congregation of the Index. This document was most artfully drawn up; too cowardly to mention Galileo by name, as he knew him to have friends at the Vatican, he denounces the Galileists in general, "who maintain agreeably to the doctrine of this man Ipernic, or whatever they call him, that the earth moves and the heavens stand still."

"All the fathers," he goes on, "of this devout convent of San Marco find many passages in this letter which are suspicious and presumptuous, as when it says (1) that many expressions of Holy Scripture are indefinite; (2) that in discussions about natural phenomena the lowest place must be assigned to them; (3) that the commentators have often been mistaken in their interpretations; (4) that the Scriptures should not be mixed with anything but matters of religion; (5) that in nature philosophical and astronomical evidence is of more value than Holy Writ; and finally (6) that when Joshua commanded the sun to stand still we must only understand that the command was addressed to the *primum mobile,* and not to the sun"; and so on, and so on.

In consequence of this denunciation the Holy Office instituted a secret enquiry. As Lorini had only sent a copy of the letter to Castelli, the Inquisition took steps to gain possession of the original, written and signed by Galileo, and wrote (26th February 1615) to the Archbishop of Pisa to procure the document "in a skilful manner." But in this they were foiled; Castelli no longer possessed the letter, having returned it to Galileo, and the latter, on being asked for it again, supplied only a copy, unsigned and undated, and which, moreover, he strictly enjoined Castelli not to let out of his hands. Castelli accordingly read the letter to Archbishop Bonciani (in presence of several canons of his cathedral), who concealed as well as he could his annoyance at the failure of his scheme. So, Cardinal Mellini had to be content with Lorini's copy of the incriminated letter, which was submitted to the Councillor of the Holy Office for his opinion. His report was a colourless one: some passages were objectionable but, although at first sight they looked ill, they were capable of being taken in a good sense, and on the whole the document was not of that nature that it could be said to deviate from Catholic doctrine.

Meanwhile, Caccini, who had been transferred to Rome, was summoned as one specially well-informed about Galileo's teachings! His evidence was a repetition of his sermon, mixed up with all the tittle-tattle of Florence. After denouncing Galileo and his works as Godless, he concluded, "At any rate he is suspicious in religious matters, because he belongs to a certain Accademia dei Lincei, and corresponds with the Godless Fra Paolo Sarpi at Venice, and with many Germans."

Galileo appears to have known nothing accurately of these proceedings. He could only learn that some Dominican monks were making use of his letter to Castelli in order to effect the condemnation of the Copernican doctrines. Fearing that the copy on which they relied at Rome might have been tampered with, he sent a correct copy, on 16th February 1615, to his friend and old pupil, Mgr. Piero Dini, with a request that he would forward it to Father Griemberger, and perhaps also to Cardinal Bellarmine. He added that he had written to Castelli *currente calamo,* and that he had since made many researches into the matters therein discussed, which he was embodying in a larger treatise —a copy of which when finished he would send. Mgr. Dini in reply, and apparently as the result of a conference with Cardinals del Monte and Bellarmine, urged Galileo to hurry on his writing, but at the same time advised caution, to avoid theology, and to confine himself to mathematics and physics. "Write freely," he said, "but be careful to keep outside the sacristy." Mgr. Ciampoli, another (Florentine) disciple then residing in Rome, writing 28th February, conveyed the assurance of Cardinal Maffeo Barberini that he had Galileo's interests at heart, and that it would be more prudent not to go beyond his role of Professor of Mathematics, as the theologians claimed to have the sole right of explaining the Scriptures. Prince Cesi gave him much the same advice, and added his belief that any deliberate formulation of the Copernican doctrine would only end in its prohibition, and the suppression of his book. This excellent advice unfortunately came too late, for, as we see, the Inquisition was already moving; and even if Galileo had never written his letter to the Grand Duchess, or had written it in the way suggested, his enemies already had grounds enough to go upon.

Although down to the end of June 1615, Galileo received reassuring letters from Rome, that there was nothing in progress at which he need be at all disconcerted, yet, so well did the Inquisition keep its secret, it was all the time laying its plans to entrap him.

MEANWHILE, about the middle of 1615, Galileo completed his great apologetic treatise in the form of a letter to the Dowager Grand Duchess Christina, in which he defines at great length his position as a natural philosopher, and as a sincere member of the Church of Rome. It is, naturally, an amplification of his letter to Castelli already quoted, and need not, therefore, be more fully noticed. The pith of it is contained in the saying of Cardinal Baronius, which the author quotes: "The Holy Spirit intended to teach us in the Bible how to go to heaven, not how the heavens go."

The concluding passages may be quoted as follows:

> Again, to command professors of astronomy that they must themselves see to confuting their own observations and demonstrations is to ask the impossible, for it is not only to command them not to see what they do see, and not to understand what they do understand, but to seek for and to find the contrary. I would entreat these wise and prudent Fathers to consider diligently the difference between opinionative and demonstrative doctrines, to the end that they may assure themselves that it is not in the power of professors of demonstrative Sciences to change their opinions at pleasure, and adopt first one side and then the other; that there is a great difference between ordering a mathematician, or a philosopher, as to what opinion to hold, and doing the same with a merchant, or a lawyer, since demonstrated conclusions touching things of nature and of the heavens cannot be changed with the same facility as opinions touching what is lawful, or not, in a contract, bargain, or bill of exchange.
>
> Therefore, let such people apply themselves to the study of the arguments of Copernicus and others, and leave the condemning of them as erroneous and heretical to whom it belongeth. Yet, as to this latter, they must not hope to find such rash and precipitate determinations in the wary Holy Fathers, or in the absolute wisdom of him who cannot err (the Pope), as those into which they suffer themselves to be hurried by some particular affection or interest of their own. In these, and such like opinions which are not directly articles of faith, certainly no man doubts that his Holiness hath always an absolute power of admitting or condemning them; but it is not in the power of any

creature to make them to be true or false, otherwise than as, in fact, they are.

This letter, notwithstanding its moderate and even deprecating tone, was, of course, a fresh weapon against him, and his enemies denounced him more noisily than ever. Ominous reports began to circulate, but still Galileo could not learn any thing more definite than that something was brewing against him, and that the Copernican theory would probably be interdicted. Thinking that he could best combat these intrigues by going to Rome, he set out early in December 1615, provided with cordial letters from the Grand Duke to his Ambassador Guicciardini, to Cardinals del Monte and Orsini, and others. His enemies spread about that this step did not originate with himself, but was the result of a direct citation. This, however, is abundantly disproved by the Grand Duke's letters in which he distinctly says that Galileo goes to Rome of his own accord *(spontaneamente)*.

~~CHRIST AND THE TRIBUTE MONEY~~

GALILEO AND THE APPARENT CONFLICT BETWEEN SCIENCE AND RELIGION

Render therefore into Scientific language the things that concern Scientists, and into religious language the things that concern everyone else.

HIS RECEPTION by the authorities seemed to be cordial, but he quickly discovered that a zealous agitation was going on, not only against his teachings but against himself. However, he was confident that he should very soon destroy the traps of his enemies, and justify himself in a way that would bring discredit upon themselves. Judging from a letter of some days later, 23rd January 1616, to Secretary Picchena, he had not found his defence so easy as he anticipated. He says:

> My business is far more difficult and takes much longer, owing to untoward circumstances, than the nature of it would require. I cannot communicate directly with those persons with whom I have to negotiate—partly to avoid doing injury to any of my friends, and partly because they cannot communicate anything to me without running the risk of grave censure. So I am compelled with much pains and caution to seek out third persons who, without even knowing my object, may serve as mediators with the principals. ... I have also to set down some points in writing, and to cause that they shall come privately into the hands of those whom I wish to see them.

At length Galileo succeeded, as he thought, in freeing himself personally from all accusations and in refuting the slanders of Caccini and his confederates. The monk paid him a long visit, humbly begged pardon for his previous conduct, and offered any satisfaction in his power. But Galileo rightly gauged his sincerity, in a letter to Picchena of 20th February, in which he says: "I perceived not only his great ignorance, but that he has a mind void of charity and full of venom."

Lorini excused himself in a still more contemptible manner. He coolly admitted that he knew nothing and wanted to know nothing of the merits of the controversy. He only acted "for the sake of saying something, lest men should think that the Dominican Fathers were asleep or dead."

But by adjusting his own difficulties Galileo had performed only half the task he set himself. The grander part of it, viz. the preservation of the Copernican doctrines from the interdict of the Church, had yet to be accomplished.

> My business, so far as it concerns myself, is completed. All the exalted personages who have been conducting it have told me so plainly and in a most obliging manner, and have assured me that people are fully convinced of my uprightness and honour, and of the devilish malice and injustice of my persecutors. As far, therefore, as this matter is concerned I might return home without delay; but there is a question connected with my case, which does not concern myself alone, but all those who during the last eighty years have advocated, in printed works and private letters, in public lectures and private conversations, a certain theory not unknown to your Excellency, on which they are now proposing to pronounce judgment. In the hope that my assistance may be of use in this matter, so far as a knowledge of those truths is concerned which are proved by the Science to which I have devoted myself, I as a zealous and Catholic Christian neither can nor ought to withhold that assistance which my knowledge affords; and this part of the business keeps me fully employed.

As illustrating his manner of assisting the Copernican cause, the following letter from Antonio Querengo to Cardinal d'Este, 20th January 1616, is interesting.

> Your Reverence would be delighted with Galileo if you heard him holding forth, as he often does, in the midst of fifteen or twenty,

> all violently attacking him, sometimes in one house, sometimes in another. But he is armed after such fashion that he laughs all of them to scorn; and even if the novelty of his opinions prevents entire persuasion, at least it convicts of emptiness most of the arguments with which his adversaries endeavour to overwhelm him. He was particularly admirable on Monday last, in the house of Signor Federigo Ghisilieri; and what especially pleased me was that, before replying to the contrary arguments, he amplified and enforced them with new grounds of great plausibility, so as to leave his adversaries in a more ridiculous plight when he afterwards overturned them all.

This entering the lists in the cause of Copernicanism was, to say the least, magnanimous, and Galileo was entitled as no other was to appear as an advocate; but unfortunately his warm and, perhaps, too solicitous efforts in the cause had a result precisely opposite to the one he intended. The fact is, Galileo's too sanguine temperament deceived him woefully as to the true course of these transactions, as we shall also find to be the case in his second and more terrible encounter with the Holy Office; for, a few days after he thought his own affairs so satisfactorily settled that he could if he chose return to Florence in triumph, a bolt shot out of the blue which paralysed him, and spread consternation among his friends. Without any warning (for such was the manner of the Inquisition), on 19th February, the Qualifiers, or official experts, of the Holy Office were called on for an opinion on the following propositions extracted from Galileo's work on the sun-spots:

1. The sun is the centre of the world, and, therefore, immovable from its place.
2. The earth is not the centre of the world, and is not immovable, but moves, and also with a diurnal motion.

On the 24th February the Qualifiers reported:

1. The first proposition is unanimously declared to be false and absurd philosophically, and formally heretical, inasmuch as it expressly contradicts the doctrines of Holy Scripture in many passages, both if taken in their literal meaning and according to the interpretation of the Holy Fathers and learned theologians.
2. The second proposition is declared unanimously to deserve the like censure (as the first) in philosophy, and, as regards its theological aspect, to be at least erroneous in faith.

On 25th February, Cardinal Mellini reported this opinion to the Holy Office, and, as the result, Cardinal Bellarmine was directed "to summon before him the said Galileo, and admonish him to abandon the said opinion; and in case of refusal the Commissary is to intimate to him, before a notary and witnesses, a command to abstain altogether from teaching or defending the said opinion, and even from discussing it; and if he do not acquiesce therein he is to be imprisoned."

This admonition was administered on the next day, as appears from the following minute:

Friday, the 26th.

At the palace, the usual residence of the Lord Cardinal Bellarmine, the said Galileo, having been summoned and brought before the said Lord Cardinal, was, in presence of the Most Revd. Michelangelo Seghizzi, of the order of preachers. Commissary-General of the Holy Office, warned by the said Lord Cardinal of the error of the aforesaid opinion and admonished to abandon it. And immediately thereafter before me and before witnesses, the Lord Cardinal Bellarmine being still present, the said Galileo was by the said Commissary commanded and enjoined, in the name of his Holiness the Pope, and the whole Congregation of the Holy Office, to relinquish altogether the said opinion, that the sun is the centre of the world, and immovable, and that the earth moves; nor henceforth to hold, teach, or defend it in any way whatsoever, verbally or in writing; otherwise proceedings would be taken against him in the Holy Office; which injunction the said Galileo acquiesced in and promised to obey.

Done at Rome, in the palace aforesaid, in the presence of Badino Nores, of Nicosia, in the Kingdom of Cyprus, and Augustino Mongardo, from a place in the abbacy of Rose, in the diocese of Politianeti, inmates of the said Cardinal's house, witnesses.

Here the process ended so far as Galileo was concerned, but it was followed up a few days later by an edict prohibiting and suspending certain writings, amongst them the book of Copernicus. This decree runs as follows:

March 3, 1616.

The Lord Cardinal Bellarmine having reported that Galileo Galilei, mathematician, had, in terms of the order of the Holy Congregation, been admonished to abandon the opinion he has hitherto held, and had acquiesced therein; and the decree of the Congregation of the Index having been presented, prohibiting and suspending respectively the writings of Nicolas Copernicus on *The Revolutions of the Celestial Orbs;* of Diego di Zuniga *On Job;* and of Paolo Antonio Foscarini, Carmelite friar; his Holiness has ordered this edict of prohibition and suspension respectively to be published as follows:

March 5, 1616 [after preamble].

And whereas it has also come to the knowledge of the said Congregation that the Pythagorean doctrine (which is false and altogether opposed to Holy Scripture) of the motion of the earth and the quiescence of the sun, which is also taught by Nicolas Copernicus in his *Revolutions of the Celestial Orbs,* and by Diego di Zuniga in his book *On Job,* is now being spread abroad and accepted by many—as may be seen from a certain letter of a Carmelite father, wherein he attempts to show that the aforesaid doctrine is consonant with truth and is not opposed to Holy Scripture. Therefore, in order that this opinion may not insinuate itself any further to the prejudice of Catholic truth, the Holy Congregation has decreed that the said works of Nicolas Copernicus and Diego di Zuniga be suspended until they be corrected; that the book of the Carmelite, Foscarini, be altogether prohibited and condemned; and that all other books in which the same doctrine is taught be likewise prohibited.

It should be noted that the book of Copernicus was only suspended until corrected. These "corrections," which were of the most trivial character, were carried out under the supervision of Cardinal Gaetani in 1620, after which it would be allowable to read the book, and its doctrines could be held by the faithful, "*ex hypothesi,* and without affirming anything."

GALILEO REMAINED in Rome for nearly three months after the promulgation of the decree of 5th March. His enemies spread the report that he had been obliged to recant formally and to abjure his opinions. It would be easy to show his friends how grossly exaggerated was this report; but in order the better to confute these calumnies, and to guard against them in the future, he obtained a paper from Cardinal Bellarmine, stating the facts as follows:

> We, Robert Cardinal Bellarmine, having heard that it is calumniously reported that Signor Galileo Galilei has in our hand abjured, and has also been punished with salutary penance, and being requested to state the truth as to this, declare that the said Signor Galileo Galilei has not abjured, either in our hand or the hand of any other person here in Rome, or anywhere else, so far as we know, any opinion or doctrine held by him; neither has any salutary penance been imposed upon him, but only the declaration made by the Holy Father, and published by the sacred Congregation of the Index, has been intimated to him, wherein it is set forth that the doctrine attributed to Copernicus—that the earth moves round the sun, and that the sun is stationary in the centre of the world, and does not move from east to west—is contrary to the Holy Scriptures, and therefore cannot be defended or held.
>
> In witness whereof we have written and subscribed these presents with our hand this 26th day of May 1616.

Galileo having taken his admonition was admitted to an audience of the Pope, Paul V, on 11th March, which lasted three-quarters of an hour. He took occasion to refer to the events just concluded, and expressed his fears of never-ending persecution; but Paul V consoled him with the assurance that he need have no fear; that he was held in so much esteem by himself and the whole Congregation, that they would not listen to any calumnies; and that, as long as he occupied the chair of St Peter, Galileo might rest assured that he was safe from all danger.

On the eve of Galileo's departure from Rome, Cardinal del Monte wrote to the Grand Duke "to bear witness that he was leaving with the best reputation, and the approval of all who have had transactions with him, for it has been made manifest how unjust have been the calumnies of his enemies." However, the Grand Duke's Ambassador, Guicciardini, told a different story, and, perhaps, a truer one. All along he had tried to get Galileo recalled to Florence; his last letter of 13th May may be cited as a specimen of his previous despatches.

"Galileo," he says, "seems disposed to emulate the monks in obstinacy, and to contend with personages whom you cannot attack without ruin to yourself. It may any moment be heard in Florence that he has stumbled into some abyss or other. However, the heat will probably drive him from Rome before long, and that will be the best thing that can happen to him."

Cosimo II, alarmed by these gloomy reports of his ambassador, at last issued orders for Galileo's return, and on 23rd May Picchena wrote:

> You have had enough of monkish persecutions and ought to know by this what the flavour of them is. His Highness fears that your longer stay in Rome may involve you in fresh difficulties, and would therefore be glad if (as you have so far come honourably out of the affair) you would not tease the sleeping dog any more, and would return here as soon as possible. There are rumours flying about which we do not like, and the monks are all powerful. I, your servant, must not fail to warn you, and to inform you, as in duty bound, of the wishes of our Master, wherewith I kiss your hand.

Galileo complied with this order with the least possible delay, and set out on his homeward journey on 4th June 1616.

8

GALILEO IN FLORENCE

1617–1624

For a long time after his return to Florence in June 1616, Galileo's health had been very indifferent. A complication of his old maladies, aggravated by long fits of hypochondria, left him little time or taste for work, and sadly interrupted his correspondence. His old friend Sagredo often advised him to take his ease and be content with the laurels he had already won. "Philosophise," he used to say, "comfortably in your bed and leave the stars alone. Let fools be fools, let the ignorant plume themselves on their ignorance. Why should you court martyrdom for the sake of winning them from their folly? It is not given to everyone to be of the number of the elect. I believe the universe was made for my service, not I for the universe. Live as I do and you will be happy." This, indeed, was the burden of Sagredo's letters from the time his friend quitted Padua in 1610 up to his own death on 5th March 1620.

But Galileo had not the epicurean temperament of his friend. Speculation and experiment were as necessary to him as food and air; yet from these he was now practically debarred by the Papal prohibition of 26th February. How could he resume his old work, or begin any new one? The Copernican theory of the world was the basis of all his work. Its acceptance and application were forbidden him, and the general permission to employ it as a working hypothesis was a mockery. Can we wonder then at the fits of melancholy which frequently oppressed him? Yet inaction to a man of his temperament was intolerable; he must be doing something; he was fond of gardening, but he could not be always digging, and weeding, and pruning. He must be doing something else; he could not write for publication, but he could at least commune with his intimate friends; and so, for some years he poured out his thoughts in long private letters, copies of which were circulated amongst the learned men of Europe. Unfortunately, few of these are now extant; but amongst these few there is one which deserves notice, as it shows how intolerable the writer felt his position to be. It is also interesting as a specimen of the keen sarcasm of which he was a master. During his stay at Rome, as we have seen, he wrote a treatise on the tides. Now, on 23rd May 1618, he sent a copy of this to the Archduke Leopold of Austria; but as since it was written the proceedings culminating in the decree of 5th March 1616 had taken place, Galileo added the following accompaniment:

> With this I send a treatise on the causes of the tides, which I wrote more than two years ago, at the suggestion of his Reverence Cardinal Orsini in Rome, and at the time when the theologians were thinking of prohibiting the book of Copernicus and the doctrine enunciated therein of the motion of the earth, which I held to be true until it pleased those gentlemen to prohibit the work and to declare that opinion to be false and contrary to Scripture. Now, knowing as I do that it behooves us to obey the decisions of the authorities and to believe them, since they are guided by a higher insight than any to which my humble mind can of itself attain, I consider this treatise which I send you merely to be a poetical conceit or a dream, and desire that your Highness may take it as such, inasmuch as it is based on the double motion of the earth, and, indeed, contains one of the arguments which I have adduced in confirmation of it. But even poets sometimes attach a value to one or other of their

fantasies, and I likewise attach some value to this fancy of mine. Now, having written this treatise and having shown it to the Cardinal above mentioned and a few others, I have also let a few exalted personages have copies, in order that, in case any one not belonging to our Church should try to appropriate my curious fancy (as has happened to me with many of my discoveries), these personages, being above all suspicion, may be able to bear witness that it was I who first dreamed of this chimera. What I now send is but a fugitive performance. It was written in haste and in the expectation that the work of Copernicus would not be condemned as erroneous eighty years after its publication. I had intended at my convenience and in quiet to have gone more particularly into this subject, to have added more proofs, to have arranged the whole anew, and to have put it into a better form. But a voice from heaven has aroused me and dissolved all my confused and tangled fantasies in mist! May therefore your Highness graciously accept it, ill-arranged as it is; and if Divine love ever grants that I may be in a position to exert myself a little, your Highness may expect something more solid and real from me.

IN AUGUST 1618, three comets appeared in the heavens, and the brilliant one in the constellation of the Scorpion—one of the most splendid of modern times—especially attracted the attention of astronomers. Although this was visible until January 1619, Galileo had little opportunity of observing it, as he was confined to bed nearly the whole time by severe illness. However, we may suppose with Viviani that he was able to observe a little, but, certainly, he reflected much, and, as soon as he was able, he communicated his views to his friends, amongst others to the above-mentioned Archduke Leopold, who, being in Florence on a visit, came to see him.

His views were published about the middle of 1619, through the medium of Mario Guiducci, a Florentine disciple. He did not consider comets to be really heavenly bodies, analogous to the planets as was currently supposed, but only atmospheric phenomena—columns of vapour which rise from the earth to a great height, far beyond the moon, and become temporarily visible by refraction of the sun's rays. In fact, he classed comets in the same category as rainbows and mock suns. Referring to some proposed parallax measurements of the comets, he pointed out the difference in this respect between a fixed object, the distance of which may be calculated by two angular observations at a known distance apart, and atmospheric appearances like rainbows which are simultaneously formed in different drops of water for each spectator, so that two observers in different places are, in fact, viewing different objects. He then warns astronomers not to engage with too much warmth in a discussion on the distance of comets before they assure themselves to which of these two classes of phenomena they are to be referred. The remark is in itself perfectly just, although the opinion which occasioned it is now known to be erroneous; but it is questionable whether the few observations which up to that time had been made upon comets were sufficient to justify the bitter censures which have been cast on Galileo on account of it. Indeed, the same opinion was held for a time by Cassini, the celebrated astronomer of the Paris Observatory, many years after Galileo, and when the Science was considerably more advanced; and even Newton did not consider it beneath notice, for in his *Principia* he takes pains to show on what grounds it is untenable.

In the course of Guiducci's essay, some opinions of the Jesuit father, Orazio Grassi (in a public discussion on the comets at the Collegio Romano), were so indiscreetly attacked as to raise the ire of the whole Jesuits' College at Rome. Grassi, under the pseudonym of Lotario Sarsi, published an onslaught on Galileo's cometary theory in a book called *The Astronomical and Philosophical Balance* (1619)—a violent pamphlet full of abuse of Galileo and his school. Friends, like Prince Cesi, and Mgrs. Ciampoli and Cesarini, now advised that the master himself should take up the fight; but ill-health and caution, owing to the troubled state of the religious and political horizons, prevented the appearance of his reply for three years. At length, in October 1622, he sent the manuscript of this celebrated work, "Il Saggiatore" (The Assayer), to Mgr. Cesarini in Rome, and for five months it passed from hand to hand among the members of the Accademia del Lincei, who examined it carefully and (with the author's consent) altered some passages which might possibly have given a handle to his enemies. The Papal Imprimatur was granted 2nd February 1623, on the report of Father Niccolo Riccardi, Examiner, who was formerly a pupil of Galileo, and of whom we shall hear a great deal in the sequel. This report was as follows:

> By command of the Master of the Palace, I have read the work 'Il Saggiatore,' and not only have I detected nothing in it which is contrary to good morals, or deviates from the Divine Truth of our religion, but I have found in it such beautiful and manifold observations on natural philosophy that I think our age will not have to boast merely of having been the inheritor of the labours of earlier philosophers, but also of having been the discoverer of many secrets of nature which they were not able to penetrate, thanks to the subtle and solid researches of the author whose contemporary I think myself happy to be.

WHILE THE WORK was in the press, an event occurred which seemed likely to produce a change for the better in Galileo's relations with Rome. On 8th July 1623, Gregory XV succumbed to age and infirmity in the second year of his pontificate; and the man who at the age of fifty-five succeeded him (elected 8th August 1623) was Cardinal Maffeo Barberini, now Pope Urban VIII. Galileo placed great hopes for the progress of Science in general, and for, at least, toleration of the Copernican doctrine, on this election; and to all appearances he was justified in doing so. Not only was Urban VIII a refreshing contrast to his immediate predecessors, who cared little for art or Science, but, as Cardinal, he had for many years entertained a great friendship for Galileo, as many letters of his still extant show. Thus, writing from Bologna, 5th June 1612, on receipt of a copy of the work on Floating Bodies, he says:

> I have received your treatise on various Scientific questions which have been raised during my stay here, and shall read them with great pleasure, both to confirm myself in my opinions which agree with yours, and to enjoy with the rest of the world the fruits of your rare intellect.

Another letter of 20th April 1613, after the work on sun-spots had appeared, may be quoted:

> Your printed letters to Welser have reached me, and are very welcome. I shall not fail to read them with pleasure again and again as they deserve. This is not a book which will be allowed to stand idly among others. It is the only one which can induce me to withdraw for a few hours from my official duties to devote myself to its perusal, and to the observation of the planets of which it treats—if the telescopes we have here are fit for it. Meanwhile I thank you very much for your remembrance of me, and beg you not to forget the high opinion which I entertain for a mind so extraordinarily gifted as yours.

The Cardinal had not confined himself to mere assurances of esteem and friendship in his letters, but had shown them in his acts. Thus, in the troublous times in 1615–16, his influence with Pope Paul V greatly helped Galileo to extricate himself from his difficulties.

In 1620, Barberini gave another and a really enthusiastic proof of his regard. He celebrated Galileo's discoveries in some elegant verses (in which astronomy was allied with morality) and sent them with the following letter, dated 28th August:

> The esteem, which I always entertain for yourself and for your great merits, has given occasion to the enclosed verses. If not worthy of you they will serve at any rate as a proof of my affection; while I purpose to add lustre to my poetry by coupling it with your renowned name. Without wasting words in further apologies, I beg you to receive with favour this small proof of my great esteem.

After much delay in the printing, "Il Saggiatore" appeared at the end of October 1623, with a dedication to Pope Urban VIII, and under the auspices of the Accademia dei Lincei. This celebrated work is a masterpiece of ingenuity, for the author not only dextrously avoids the snares laid for him by Father Grassi and his supporters, but brings defeat and ridicule upon them at every turn. All this is done in so sparkling a style, and the reasoning, and counter-refutations are so convincing, that "Il Saggiatore" deserves its reputation as a model of dialectic skill, and an ornament of classical Italian literature.

The book was a great success, and, of course, intensified the bitterness of the Jesuitical party; so much so that the General of the Order forbade, under severe penalties, the members to speak of it even among themselves. It is important to note that the Pope was delighted with it, and had it read aloud to him at table.

Early in 1625, the book was denounced anonymously to the Inquisition as a veiled defence of the Copernican doctrines, and a movement was begun to prohibit it, or, at least, to have it "corrected"; but the attempt failed, and only brought discredit upon the agitators. Father Guevara, General of the Theatines, to whom it was submitted for examination, reported most favourably of it, and went so far as to say that, even if the doctrine of the earth's motion had been maintained, it would not have appeared to him a sufficient reason for condemning the work.

On the election of Cardinal Barberini to the Papacy as Urban VIII, Galileo conceived the idea of going to Rome to offer his congratulations in person, and to use his influence with the new Pope to obtain, at least, toleration for the Copernican doctrines, now no longer opposed by the weighty influence of Cardinal Bellarmine, who had died two years before. Remembering the warmth of Barberini's letters while Cardinal, Galileo had reason to hope from a Pontiff so enlightened at least this much; while as regarded himself, he felt that he must have permission to teach the new astronomical doctrines, not merely as hypotheses but as actual truths, now or never; and according as his desire was fulfilled or not, so would his life-work be complete or incomplete.

AS WE ARE NOW APPROACHING the great crisis of Galileo's public life, it will be convenient to pause and take a glance at his family affairs. We have seen that he was never married, but that he had by the Venetian, Marina Gamba, three children—Virginia, Livia, and Vincenzio. These children were brought up with the mother in Padua, and in a separate establishment from that occupied by the father in Via Vignali.

It would appear that when Galileo quitted Padua in the autumn of 1610, he took the two girls with him to Florence, and placed them under the care of his brother-in-law, Landucci; Vincenzio, being then only four years old, was left behind with his mother until October 1612, when he too was brought to Florence. Shortly afterwards, Marina married a well-to-do man in her own sphere, one Giovanni Bartoluzzi. This step appears to have been taken with Galileo's approval, judging from the respectful and friendly tone of the only letter of Bartoluzzi (17th August 1619) found amongst Galileo's papers. It appears from this letter, and from two others written by Liceti, 31st December 1610, and Pignoria, 25th January 1613, that Galileo behaved with great liberality to Marina and her husband.

In March 1610, before his final departure from Padua, he had the intention of placing his elder daughter as a boarder in the convent of the Nunziatina, Florence; but, although all the preliminaries were settled, the project, for some unknown reason, was not carried out. What to do with the girls was now become a serious question. Their taint of birth was in painful contrast to the honoured name of his own noble family, as his mother, in her frequent and "terrible" tempers, did not fail to remind him. His means, generally as we have seen insufficient for his wants and never too large, did not hold out the hope of being able to make them independent, or to settle them suitably in marriage. In these circumstances he resolved, while they were yet young, and before they could acquire a taste for the world, to place them both in a convent for life. In November 1611, he took steps to carry out this resolution, but met with many difficulties. He did not wish that the children should be separated, but there was a strict rule against sisters taking the veil in the same convent. Then, there was the further difficulty of the girls being much under the canonical age of full sixteen—Virginia, the elder, being only eleven. However, after long negotiations, and finally through the influence of Cardinal Bandini, the necessary licences were obtained, and in October 1613 the two girls were placed in the Convent of San Matteo, Arcetri, near Florence, as a preparatory step to their novitiate and final profession. In July 1614 they were entered as novices of the Order. The Mother Abbess of the time was a sister of the Secretary Vinta, whose name has often occurred in these pages. From the first the good Lodovica Vinta took a kindly interest in the poor children; and at her suggestion the feasting usual on taking the veil was dispensed with. "It would be better," she said, "in every way for the ceremony to take place quietly, and the money will be far better employed in adding to the girls' little comforts in the convent than in regaling friends and relations." Finally, Virginia became a professed nun on 4th October 1616, under the name of Suor Maria Celeste; and Livia on 28th October 1617, under the name of Suor Arcangela; and here we shall leave them for awhile.

After the death of Filippo Salviati, on 22nd March 1614, Galileo appears to have given up the villa near Signa, and to have had no settled home of his own for the next three years. He probably had a pied-à-terre in his mother's house, where, owing to her terrible tempers and his own frequent illnesses, his lot was not a happy one.

On 15th August 1617, he rented the villa of Lorenzo Segni on the Bellosguardo Hill outside Florence, and here he lived for the next fourteen years. Perched upon a hill, it commanded most lovely views of the city and the silvery Arno at its foot, with the far-famed Fiesole beyond, and of beautiful country all round.

For nearly seven years after taking the veil we lose sight of Galileo's daughters. We left them as children of sixteen; we are now to meet them as women. Sister Arcangela, the younger, we shall not like; but Virginia, or Sister Maria Celeste, as we must henceforth call her, we shall learn to love. All we know of this charming personality is told in her letters to her father. The first is dated 10th May 1623, but Professor Favaro thinks there must have been many previous ones which are now lost. However this may be, those that remain, one hundred and twenty-four in number, show that there was a close intimacy and affection between father and daughter, and that these relations grew in intensity with the daughter's maturing years. If they did not write they must have often met, for Galileo's house at Bellosguardo was a pleasant half-hour's walk along a charming road from the convent of San Matteo. In no other way than on the supposition of a previous intimacy of a close personal kind can we account for the affection which bursts forth in the first letter we possess, and overflows through all of them to the end. This affection must have been, as indeed we shall see it was, a great comfort and consolation to Galileo, sorely tried as he was by frequent illness, by the worries and ingratitude of all the rest of his family, and by the persecution of the outside world.

9

FOURTH VISIT TO ROME—RETURN TO FLORENCE

1624–1629

Long delayed, first by illness and then by bad weather, whole tracts of country being under water, Galileo at length set out for Rome 1st April 1624. Reaching Acquasparta on 5th April, he stayed a fortnight with his friend Prince Cesi, and arrived in Rome on the 23rd, provided with a warm letter of recommendation from the Grand Duchess Christina to her son Cardinal de Medici. From Acquasparta he wrote to his daughter, telling her of the flattering reception he had met with from Prince Cesi, who was able to assure him that his presence in Rome was anxiously awaited "by great personages." On the other hand he was grieved to hear of the sudden death of Mgr. Cesarini, a warm friend whom he had both loved and honoured. This event, as Maria Celeste reminds her father (26th April), "gives food for reflection on the vanity and fallacy of all earthly hopes"; but she timidly adds, "I would not have you think that I write only to sermonise you, therefore I will say no more."

All the world of Rome was aware of the favour in which the Pope held Galileo. His old friends, therefore, received him with greater delight than ever, and his enemies dared only to clench their fists behind his back. His letters express the great satisfaction which his reception afforded him; but as regarded the object which was nearest his heart he was not so satisfied. Within six weeks he had had six long interviews with Urban VIII, had always been most affably received, and was allowed to bring forward all his arguments in support of the Copernican theory. But this was all to no purpose; the Pope listened to his arguments, but would not grant his earnest entreaties for, at least, a passive toleration of the new doctrines.

As soon as the Pope's attitude became known, Galileo's clerical friends had to be cautious, and avoided as far as possible all reference to the prohibited doctrines. One of these, Father Niccolo Riccardi, at once took his seat on the fence.

"As to the truth or falsity of the theory," wrote Galileo to Cesi on 8th June, "he accepts neither Ptolemy nor Copernicus, but quiets his soul in a very speedy manner. He sets angels to work at moving the heavenly bodies, and these make them go as they do go (however that may be) without the slightest difficulty or entanglement! Certainly this ought to be enough for us!"

Finding that his efforts to get the decree of 5th March 1616 revoked were of no avail, Galileo resolved with a heavy heart to return home, after a six weeks' stay in Rome. On the eve of his departure, the Pope loaded him with favours which must have seemed to him like mockeries. His Holiness promised him a pension for his son, and three days after sent a picture for himself; then two medals—one of gold and one of silver, and quite a number of Agnus Dei! Not content with these marks of favour, he addressed an official letter *(Breve)* to the Grand Duke, 8th June, in which, to the no small chagrin of Galileo's enemies, his Holiness not only did full justice to our philosopher's services to Science, "the fame of which will shine on earth so long as Jupiter and his satellites shine in heaven," but laid special stress on his religious sentiments:

> We have observed in him not only literary distinction, but love of religion, and all good qualities worthy of the Papal favour. When he came to congratulate us on our accession we embraced him affectionately and listened with pleasure to his learned

> demonstrations, which add fresh renown to Florentine eloquence. We desire that he should not return to his native country without receiving from our generosity manifold proofs of our favour. And that you may fully understand to what extent he is dear to us, we give this honourable testimony to his virtue and piety. And, further, we assure you that we shall thank you for any kindness that you can show him; and by imitating, or even surpassing, our fatherly liberality, you will add to our gratification.

FRUITLESS AS WAS his journey to Rome as regarded the grand object of his life—the emancipation of the Copernican theory—Galileo was yet able to do something for the advancement of Science. He improved, if he did not invent, the microscope. The principle of the telescope and the microscope is to the mathematical optician one and the same. The former is merely made to collect parallel rays from distant objects; the latter, diverging rays from near objects. The invention of the one, therefore, could hardly fail to follow immediately upon the other; and accordingly we learn that, very soon after inventing the telescope, Galileo adapted it for the examination of small objects. John Wedderburn, a Scotch student at Padua, in a defence of his master (published in 1610) against the calumnies of Martin Horky (see p. 106 *ante*), states that he heard Galileo describe in what manner he perfectly distinguishes with his telescope the organs of motion and of the senses of the smaller animals, especially in a certain insect, which has each eye covered by a rather thick membrane, which, perforated with seven holes, like the visor of a warrior, allows it sight.

"Here hast thou," he continues, "a new proof that the glass concentrating its rays enlarges the object. In other animals of the same size, and even smaller, some of which have, nevertheless, brighter eyes, these appear only double, with their eyebrows and the other adjacent parts."

In 1614, the Frenchman Tarde, Canon of the Cathedral of Sarlat (Dordogne), was travelling in Italy, and on arriving at Florence called on Galileo, whom he found ill in bed. Amongst other reports of this meeting, Tarde says:

> Galileo told me that the tube of a telescope for observing the stars is no more than 2 feet in length; but to see well objects, which are very near, and which on account of their small size are hardly visible to the naked eye, the tube must have two or three lengths. He tells me that with this long tube he has seen flies which look as big as a lamb, are covered all over with hair, and have very pointed nails, by means of which they keep themselves up and walk on glass, although hanging feet upwards, by inserting the points of their nails in the pores of the glass.

In "Il Saggiatore" there is a further reference to a telescope arranged so that one can see very near objects very distinctly, even to the most minute particles; and, finally, Viviani, in his *Vita di Galileo,* and in the laudatory inscriptions which he placed on the front of his house in Florence in 1693, records as a fact that Galileo presented a microscope to the King of Poland in 1612. All this goes to show that he was well acquainted with the uses of his invention *qua* microscope, and that he did not dwell upon them, or pursue them to greater length, is, no doubt, because his thoughts were wholly absorbed on its perfection as a telescope, and on the glorious field of astronomical discovery which it laid open to him. Certain it is, that for many years he gave the matter little attention—not until his visit to Rome of which we are now speaking, and when he found the microscope discussed as a novelty which nobody could understand.

We have seen in our account of the invention of the telescope that Jansen, the optician of Middleburg, invented a form of microscope about 1590, in which objects were seen inverted. One of these instruments he presented to the Archduke Charles Albert of Austria, who in turn gave it to Cornelius Drebbel, a Dutchman, then living in London. For many years after, the instrument was practically forgotten; but about 1621, Drebbel appears to have resumed its manufacture in London.

In the following year Jacob Kuffler, a relative of Drebbel, brought a specimen to Rome, a present from M. de Peiresc of Paris to one of the Cardinals. From the letter which accompanied it, dated 7th June 1622, we take the following passage:

> Your Lordship will receive the present letter from the hand of Signor G. Kuffler of Cologne. He will be able to show you an *occhiale* or telescope of a new invention (different from that of Galileo), with which he shows a flea as large as a cricket, and almost of the same shape, with its two arms and the other smaller legs, head, and almost all the body, covered with crusts or scales, like crickets or small shrimps. The little insects which generate in cheese become as large as flies, and are so distinctly discerned that one sees them to have very long legs, a pointed head, and every part of the body quite distinct.

Unfortunately, Kuffler died before he had time to explain the management of the instrument, and so it remained a mystery. Two years later, after many accidents and delays, two other specimens arrived, also sent by de Peiresc with brief instructions as to their use. Apparently one was little more than a magnifying glass, which it was easy to understand; but of the other and larger, consisting of two glasses, nobody in Rome could make anything, "although they had the help of mathematicians."

It was at this moment that Galileo arrived. The instrument was shown to him, and, as may be imagined, a very brief study told him not only how to use it but how to improve upon it. He at once told his friends that he had himself made a somewhat similar instrument many years previously, "which magnifies things as much as 50,000 times, so that one sees a fly as large as a hen." He quickly made some specimens, showing objects erect, which he sent to his friends, and soon his microscopes were in as great request as his telescopes. Amongst others, he sent one to Prince Cesi, on 23rd September 1624, with the following interesting letter:

I send your Excellency a little spy-glass *(occhialino)* for observing at close quarters the smallest objects, which I hope will afford you the same interest and pleasure that it does to myself. I delayed sending it because my first specimens were imperfect by reason of the difficulty in fashioning the lenses. The object is placed on a movable circle (at the base of the instrument) which can be turned in such a way as to show successive portions, a single pose being unable to show more than a small part of the whole. As the distance between the lens and the object must be precisely adjusted in order to see things that are in relief, it is necessary to bring the glass nearer to or farther from the object, according to the parts to be examined. Therefore the little tube is made adjustable on its stand or guide. The instrument should be used in a strong light, or even in full sunlight, so as to illuminate the object as much as possible.

I have examined with the greatest delight a large number of animals, amongst which the bug is most horrible, the gnat and the moth very beautiful. I have also been able to discover how the fly and other little animals are able to walk on window panes and ceilings feet upwards. But your Excellency will now have the opportunity of observing thousands of other details of the most curious kind, of which I shall be glad to have an account. In short, one may contemplate endlessly the grandeur of Nature, how subtilely she works, and with what unspeakable diligence.

P.S.— The little tube is in two pieces, so that you may lengthen it or shorten it at pleasure.

SOON AFTER HIS return to Florence, Galileo began to draw up a reply to an attack on the Copernican theory which had been addressed to him in 1616 by Francesco Ingoli, then a lawyer at Ravenna, and afterwards secretary of the Propaganda in Rome. Coming at the time of his first encounter with the Inquisition, Galileo wisely refrained from answering it then. In 1618 an answer was published from the pen of that other Corypheus of Science, Kepler, in his "Epitome of the Copernican Astronomy," but Ingoli did not consider himself beaten, and rejoined in a letter addressed to a high official of the Papal Court. Now, after the lapse of eight years, Galileo thought that, protected by the favour of Urban VIII, he might himself venture on a reply; for although there was no hope of a public revocation of the decree of 5th March 1616, he thought, and his correspondents in Rome were of the same opinion, that the prohibition would not be rigidly enforced against him.

In this defence of the Copernican theory, he professes to be actuated by a double motive. On the one hand he wishes to show that, as he had given currency to it before it was condemned by ecclesiastical authority, he had not been the expounder of an improbable or unreasonable theory. On the other hand he wishes to prove to the Protestant Copernicans in Germany that the views of their great countryman had not been rejected in Catholic Italy from a disbelief of their probability, but "from reverence for Holy Scripture, as well as zeal for religion and our holy faith." After this strange introduction, and an assurance that he had no intention of representing the forbidden doctrine as true, he proceeds with vigour to refute all Ingoli's objections.

In spite of his diplomatic preface, his friends in Rome, aware of the watchfulness of his enemies, advised him not to publish. Galileo wisely gave heed to their warnings, and so the work was only circulated in manuscript copies. However, several passages from it were brought under the notice of the Pope by Mgr. Ciampoli, and we learn from a letter of the latter to Galileo, 28th December 1625, that his Holiness highly approved them. This to Galileo and his friends was a hopeful sign. Another was the failure of the agitation undertaken (as already mentioned) in the same year against "Il Saggiatore." Father Grassi ventured, under pretext of a rejoinder to that work, to publish a fresh attack on its author full of spiteful personalities and "arguments" of the most absurd kind. Apparently Grassi, member though he was of the powerful Collegio Romano, could not find a publisher for this work *(Ratio Ponderum Librae)* in Rome, and had to bring it out in Paris in 1626, a circumstance which Galileo interpreted as another encouraging sign of the times. Again, in 1624 in a conversation on the subject with the Pope, the Cardinal Zollern (prompted by Galileo) represented that all the heretics of Europe considered the truth of the Copernican doctrine to be beyond doubt, and that, therefore, it would be necessary to be very circumspect in coming to any resolution upon it, to which his Holiness replied that the Church had not condemned it; nor was it to be condemned as heretical, but only as rash; and he added, that there was no fear of any one undertaking to prove that it must necessarily be true. These and other indications tended to confirm Galileo in the opinion that, under the pontificate of Urban VIII, the advocates of Copernicanism had little to fear, provided that the defence was so circumspectly handled as not to outrage the oft-mentioned decree of 5th March 1616.

On this assumption (unfortunately a mistaken one, as we shall see) he now, 1626, resolved to carry out the great work which he had long projected, and which, from the vast and varied knowledge it displayed, and from its sparkling and incisive style, was to meet with greater success than had ever been attained by any Scientific work. This was his *Dialogue on the Two Principal Systems of the World.*

During the next four years, 1626–29, Galileo was almost entirely engaged on the preparation of this great work. His official duties as Philosopher and First Mathematician to the Grand Duke did not take up much of his time, and his Scientific correspondence was not considerable; but his work was sadly interrupted by frequently recurring illnesses, and as much so by family troubles of all sorts which sorely tried his patience.

Towards the end of 1629, Galileo found himself face to face with yet another trouble which might have proved to be serious. He was menaced either with deprivation of his salary as Extraordinary Professor at Pisa, or with the loss of that leisure which had been the determining influence in his quitting Padua in 1610, and which he was now as anxious as ever to enjoy. Some ill-wishers at Pisa raised the question whether it was in the power of the Grand Duke to assign a salary out of the University funds to one who neither lectured nor resided there. This scruple had slept for nineteen years, so it is probable that those who now raised it reckoned on finding in young Ferdinando II a less firm supporter of Galileo than his father Cosimo II had been. But the matter did not proceed so far; the theologians and jurists, to whom the question was referred, decided that the Grand Duke had the power, but to put the matter beyond all further dispute, his Highness appointed Galileo to an equivalent post in the magistracy of the University, so that he was left undisturbed in the stipend and leisure which were now more than ever necessary in his old age and shattered health.

10

DIALOGUE ON THE TWO PRINCIPAL WORLD SYSTEMS

1630–1632

By the beginning of the year 1630, Galileo had completed his *Dialogues*, with the exception of an introduction or preface, an index, and a few finishing strokes here and there. In announcing this fact to his friends, he informed Prince Cesi that he intended going to Rome to see to the printing of the book—a step of which the Prince highly approved. The state of affairs at Rome just then seemed very favourable for this enterprise. Galileo's devoted disciple Castelli had been called from Pisa in 1624 to be mathematician to the Pope, and enjoyed great consideration with the Barberini family. This lifelong friend also approved the design and informed our philosopher (6th February) that Father Niccolo Riccardi, another old pupil and now chief censor of the press, had promised his assistance.

Another letter of 16th March contained equally encouraging news. According to this, the celebrated Dominican monk, Tommaso Campanella, had just told the Pope that, a short time before, he had tried to convert some German nobles to the Catholic faith; that they were favourably disposed until they heard of the prohibition of the Copernican theory, when they indignantly declined to have anything more to say to him. To this Urban replied: "It never was our intention, and if it had depended upon us that decree would not have been passed." In other letters Ciampoli and Castelli urged their old master to set out at once for the Papal residence, "where they were longing for him more than for a lady-love." In the face of these fresh indications of an altered, or, at least, tolerant attitude towards Science, we cannot be surprised at Galileo concluding that under Urban VIII, an infringement of the decree of 1616, in the spirit if not in the letter, such as his *Dialogues* undoubtedly were, would give no offence at the Vatican.

Filled with hope and with his manuscript complete, Galileo at length set out on 1st May in a Court litter, and travelling fast arrived in Rome on the evening of the 3rd. Furnished with a letter of introduction from the Grand Duke's chief secretary, Andrea Cioli, to the Tuscan ambassador, Francesco Niccolini, he was most hospitably received by that gentleman and lodged in the Embassy, where he quickly gained the friendship of the ambassador and his wife, which as we shall soon see was to be so useful to him.

Soon after his arrival he had a long audience of the Pope, and wrote on 18th May to Florence in high spirits: "His Holiness has begun to treat my affairs in a way that permits me to hope for a favourable result." Nevertheless, the result was anything but favourable; indeed, the toleration, to say nothing of the recognition, of the Copernican theory so ardently hoped for was as far off as ever. Urban VIII would not object to the publication, but certain conditions would have to be fulfilled. The title of the book, *Dialogues on the Flux and Reflux of the Tides,* was misleading and would have to be altered. The subject being really a discussion of the relative merits of the Copernican and Ptolemaic systems, this would have to be indicated in the title. The subject, moreover, would have to be treated from a purely hypothetical standpoint, and this fact must be clearly set forth in a preface or introduction. Then, the book must conclude with an argument which the Pope communicated to Galileo in 1624, and which his Holiness considered unanswerable. As great importance will attach itself to this argument in the sequel, we beg our readers to note it carefully. It is as follows: God is all-powerful; all things are therefore possible to Him; ergo the tides cannot be adduced as a necessary proof of the double motion of the earth without limiting God's omnipotence—which is absurd.

Rather than forego the publication of a work towards which he had laboured and thought for over thirty years, Galileo consented to these conditions. Doubtless, he felt that such minds as were capable of following his reasoning in favour of Copernicanism would no more be prejudiced by the hypothetical warning in the introduction than by the unanswerable argument of the conclusion.

Meanwhile, the manuscript was submitted to Father Riccardi, the Papal censor, and by him passed on to his assistant Father Raffaele Visconti, who carefully went through it, altered many passages, and finally approved the work thus revised. By this time the middle of June had arrived, and Galileo was anxious to leave Rome before the great heat set in. Riccardi read over the manuscript once more and then granted his permission for the printing of the work in Rome. Thus, by the end of June 1630, Galileo was back in Florence with his manuscript duly revised and corrected, and with the ecclesiastical Imprimatur for its publication in Rome, on the understanding that a preface and conclusion were added in accordance with the Papal wishes. But events were now at hand which long delayed Galileo's ardent desire to see his work speedily given to the world, and which involved complications afterwards taken advantage of by his watchful enemies.

SOON AFTER HIS return from Rome, Galileo wrote to his daughter, and from her reply of 21st June, we learn that he was ill again:

> Just as I was thinking of sending you a long lamentation because of your never coming near us, I received your most loving letter which shut my mouth completely. We were truly grieved to hear of your being ill; but really, after making a journey at this time of the year, and with the plague everywhere, I do not see how it could be otherwise. I am astonished to hear of your going into

> Florence every day. Pray take a few days' rest; do not even come to see us. We would prefer you kept well to the pleasure of your company.

The plague, already rife within the city walls, now began to spread to the suburbs. Even the fashionable Bellosguardo, whose reputation for salubrity equalled the beauty of its situation, was not spared. One of Galileo's own household, a glass-blower, was taken off early in October; and soon after, his son Vincenzio, seized by a panic, fled with his wife to Prato, leaving his invalid father alone, and his seven-month-old baby out at nurse in the neighbourhood of the villa!

Early in the summer of 1631, feeling age and infirmity creeping surely over him, Galileo began to think of a change of residence from Bellosguardo to the neighbourhood of the convent, where he would be able to enjoy more often his daughter's society. Maria Celeste's letters show how eager she was to hear of a house which would combine vicinity to the convent with a good situation, and a rent suitable to her father's much-drained purse. House-hunting in the neighbourhood of Florence was not then the easy work it is nowadays, such villas as existed being mostly occupied by their owners. Vincenzio, who appears to have got over his sulks and fear of the plague, was back again with his father, and helped in search for a house. Maria Celeste heard of two or three, but there was something against them all, and it was not until August that she heard of one which she considered suitable in every way. Writing on the 12th, she says:

> I am so anxious to have you in the neighbourhood that I am constantly enquiring if there is any place near here to let. I have just heard of a villa belonging to Signor Esau Martellini which is situated on the Piano de Giullari, and bounds our garden. I write at once to tell you that you may see if it be to your liking. I should be glad indeed if it were, as then I should not be obliged to remain so long without news of you as is the case at present.

Shortly afterwards, Dame Piera (Galileo's old housekeeper) going to the convent with a basket of provisions, rejoiced the daughter's heart by telling her that there was every prospect of the villa Martellini being taken. As Maria Celeste's last letter addressed to Bellosguardo is dated 30th August, it would seem that very soon after, Galileo took up his abode in the village of Arcetri in Martellini's villa, then called "Il Giojello" (the jewel), and now known as Villa Galileo. There, not five minutes' walk from the convent (indeed the grounds of the two houses adjoined), he was able to have daily intercourse with his daughters, and Maria Celeste no longer found difficulty in procuring a messenger if necessary to send affectionate enquiries about his health, or little presents from the still-room and pharmacy.

All this time, from soon after his return from Rome, Galileo was tormented with the obstacles and delays which he encountered in the printing of his *Dialogues*. It would detain us too long and be little profitable to set out these complications in detail; but they are so well summarised by Galileo himself in a letter of 7th March 1631, to Chief Secretary Cioli, that we venture to reproduce it as follows:

> As your Lordship knows, I went to Rome for the purpose of getting permission to publish my *Dialogues*, and to this end I put them in the hands of the Master of the Sacred Palace, who committed them to the care of his colleague, Father Raffaele Visconti, that he may examine them with the greatest attention, and note any doubtful matter or any conceit of imagination requiring correction, which (at my own request also) he did most thoroughly. And when I entreated the Rev. Master to grant me the required licence, his Reverence signified his wish to read the whole manuscript through once more. This was done, after

which he returned the book with the licence signed with his own hand; whereupon I, having been in Rome for two months, returned to Florence, intending to send back the book (as soon as I had added the dedication or preface, the conclusion, and a few other necessary things) to the illustrious Prince Cesi, President of the Lyncean Academy, who had always superintended the printing of my works. But owing to the death of this Prince [on 2nd August 1630] and the interruption of communications [by the plague], I was hindered from printing the work in Rome, and decided on having it done here. I had arranged matters with an able printer and publisher, and had procured the licence of the Rev. Vicar, and of the Inquisitor, and also of the illustrious Signor Niccolo Antella. I informed the Rev. Master of the Palace of all that had taken place, and of the impediments in the way of the printing in Rome. Whereupon he informed me, through our Ambassador, that he wished to have another look at the book, and that I was to send him a copy. On this I came to you, as you know, to ask if it were possible to send such a large volume to Rome with security, and you replied certainly not, and that letters were hardly safe. On this I wrote again, stating the impediments, and offering to send the preface and the end of the book, to which the superior authorities might add if they saw fit, or take away, or add notes of explanation; for I myself do not refuse to call these thoughts of mine chimeras, dreams, paralogisms, and vain imaginations, submitting the whole to the absolute wisdom of my superiors. As to the further revision of the body of the work, I suggested it might be done here by some person named by the Rev. Father. He was content that it should be so, and accordingly I sent him the preface and the end, and he authorised Father Jacinto Stefani, Counsellor of the Inquisition in Florence, to revise the work. This he did with the greatest care, observing even the minutest points which neither to him nor to my most malignant adversary could give the slightest umbrage. Indeed, the Rev. Father declared that the reading of my book had drawn tears from him more than once, when he saw with what humility and reverent submission I deferred to the authority of my superiors. And he declares, as do all who have read the book, that I ought to be entreated to publish it, instead of being hindered in so many ways, of which I need not here adduce examples.

Weeks and months ago I heard from Father Castelli that he had often met the Rev. Master who had given him to understand that he was going to send back the preface and the end, arranged to his entire satisfaction; but this has not yet been done. The papers have been thrown aside into some corner, and my life is wasting away, and I am in continual trouble. I went into town yesterday at my Serene Master's command, to see the designs for the façade of the Cathedral, and also wishing to avail myself of his kindness, so that, taking counsel with your Excellency, some means may be found for making the Rev. Master explain himself —as that the Ambassador be instructed to signify his Highness's desire for a termination of this weary business, and to let him know what sort of man his Highness has for a servant. But so exceedingly troubled was I that I could neither speak to his Highness nor look at the designs. Just now a messenger from Court has come to know how I am, and truly I am in such a state that I should not have risen from my bed had I not wished so particularly to tell your Lordship of this business, and to beg you to do for me that which I was unable to do yesterday, and to take the matter into your own hands, so that I may, while life yet remains, see what result I may expect from all my long and heavy labours. I send this by the hand of the Court messenger, and shall await your reply through Signor Geri Bocchineri. And since his Highness is so anxious to learn the state of my health I beg you to tell him that I should be pretty well in body, were I not so afflicted in mind.

The ambassador, Niccolini, was instructed to act in accordance with Galileo's wishes, and after more months of vexatious objections and delays, and not until, as Niccolini says, "formally pulled by the hair," Riccardi sent back the preface and end, now quite in order, to the censor in Florence. In his covering letter of 19th July 1631, he says:

> Conformably to the orders of our Lord (the Pope) respecting Signor Galileo's book, besides what I wrote to your Reverence concerning the body of the work, I send you the preface, with liberty to the author to alter or embellish as to the wording, so as the substance is preserved. The end may be treated in the same way.

By early autumn, and after a second revision of the whole work by Father Stefani in accordance with precise instructions from Rome, all the conditions of the censorship were finally complied with, and permission to print the book in Florence was issued in due form of Imprimatur. In all these annoying hindrances Galileo had a foretaste of the persecution which was to be his lot for the rest of his life.

WE SHALL DEVOTE the rest of this chapter to giving an idea of the plan and style of the work, and a résumé of its contents. The book, which is dedicated to Ferdinando II, bears the following title, unusually long in an age of long titles:—*Dialogue of Galileo Galilei, Lyncean, Mathematician Extraordinary of the University of Pisa, Philosopher and First Mathematician of the Most Serene Grand Duke of Tuscany; where in meetings of four days are discussed the Two Principal Systems of the World, indeterminately proposing the Philosophical and Natural arguments, as well on one side as on the other.* It is written in Italian, and in a style adapted not for the learned alone, but intelligible and attractive to everyone of ordinary education. His reason for writing in Italian instead of in Latin—the usual vehicle of philosophical subjects—is characteristic.

"I write in Italian," he says, "because I wish everyone to be able to read what I say. I see young men brought together indiscriminately to study to become physicians, philosophers, etc., who although furnished, as Ruzzante might say, with a decent set of brains, yet being unable to understand things written in gibberish, assume that in these crabbed folios there must be some grand hocus-pocus of logic and philosophy much too high up for them to jump at. I want such people to know that as Nature has given eyes to them just as well as to philosophers for the purpose of seeing her works, so she has given them brains for examining and understanding them."

The dialogue is carried on by three interlocutors, two of whom adduce the Scientific reasons for the double motion of the earth, while the third honestly tries to defend the opinions of the Ptolemaic and Aristotelian schools. Galileo gave to the defenders of the Copernican doctrine the names of two of his warmest friends, both long dead—Filippo Salviati of Florence (died 1614), and Gio Francesco Sagredo of Venice (died 1620). Salviati is the special advocate of the Copernican doctrines; Sagredo is witty, impartial, and open to conviction, a half convert, but an acute and ingenious one. To him are allotted the objections which seem to have some real force, as well as lively illustrations and digressions which would be inconsistent with the gravity of Salviati's character. Simplicio, a name borrowed from the noted Sicilian commentator of Aristotle who wrote in the sixth century, is of course a confirmed Ptolemaic and Aristotelian, and produces successively all the Scientific arguments of the peripatetic school; and as these fail to convince, he has recourse to all the arts of sophistry. Placed between the wit and the philosopher, it may be guessed that his case fares badly; in fact, he is chaffed and confuted at every turn, so that no unbiassed reader can fail to perceive the superiority of the modern theory; and as Galileo puts into the mouth of Simplicio not only every possible argument in favour of his case, but also every possible objection to the other side, this superiority is made to appear all the more striking.

The condition that the Copernican doctrine is only to be treated as an hypothesis is ostensibly complied with. If Salviati or Sagredo show the untenableness of some Ptolemaic axiom, or add a stone to the

Copernican structure, a remark is interpolated by one or the other to weaken the effect. When, for instance, it is said that the final decision in the controversy rests neither with mathematics and physics, nor with logic and philosophy, but with "a higher insight"; or when Salviati repeatedly asserts that he does not wish to maintain the Copernican doctrines as true, and uses the qualifying word "possible," or speaks of them as "fantasies" and "most vain chimeras," the reader cannot fail to see that these reservations, which always occur at critical moments, are made with the purpose of appeasing the censors.

When we remember its history, we cannot be surprised that the preface or introduction has no logical agreement with the contents of the Dialogue. It is addressed "To the Discreet Reader," and runs as follows:

> Some years ago a salutary edict was promulgated at Rome, which, in order to obviate the perilous scandals of the present age, enjoined an opportune silence on the Pythagorean opinion of the earth's motion. Some were not wanting who rashly asserted that this decree originated, not in a judicious examination, but in ill-informed passion; and complaints were heard that counsellors totally inexperienced in astronomical observations ought not by hasty prohibitions to clip the wings of speculative minds. My zeal could not keep silence when I heard these rash lamentations, and I thought it proper, as being fully informed with regard to that most prudent edict, to appear publicly as a witness of the actual truth. I happened at that time to be in Rome; I was admitted to the audiences, and enjoyed the approbation of the most eminent prelates of that Court; nor did the publication of the aforesaid decree occur without my receiving some prior intimation of it. Wherefore it is my intention in this present work to show to foreign nations that as much of this matter is known in Italy (and particularly in Rome) as ultramontane diligence can ever have formed any notion of, and (collecting together all my own speculations on the Copernican system) to show them that the

> knowledge of all these preceded the Roman censures, and that from this country proceed not only dogmas for the salvation of the soul, but also ingenious discoveries for the gratification of the understanding. With this object I have taken up in the dialogue the Copernican side of the question, treating it as a pure mathematical hypothesis, and endeavouring in every artificial manner to represent it as having the advantage, not over the opinion of the stability of the earth absolutely, but over it as taught and defended by some who profess to be peripatetics, but retain only the name, and are content, without improvement, to worship shadows, not philosophising with their own reason, but only from the recollection of four principles imperfectly understood.

The conclusion agrees no better than the preface with the body of the work. At the end of the fourth day, which is almost wholly taken up with the question of the tides, comes naturally the Pope's "unanswerable" argument of 1624 (p. 230 ante). Salviati treats it accordingly:

> It is an admirable and truly angelic argument, and perfectly in accord with that which, coming from God Himself, permits us to discuss the constitution of the world—doubtless with the view of preventing (by exercising them) the diminution and enfeeblement of our intellectual faculties, while withholding from us the power of fully comprehending the works of His hands. May this exercise (permitted and ordained by God) enable us to see and admire His greatness, which is all the more necessary since we shall never be able to penetrate the depths of His infinite wisdom.

Sagredo then says:

> Let this reflection be a fitting conclusion to our four days' discussion. And now, if Salviati desires some repose, our curiosity will concede the delay, but only on condition that at his earliest convenience he will satisfy us as to the problems reserved for future meetings. For myself, I am extremely anxious to hear his exposition of the elements of the new Science of local motions, natural and violent, as elaborated by him.

AS REGARDS the contents of the *Dialogues*, we can only give an outline. Salviati opens the conference by defining its object, which is to examine all the physical arguments evoked for and against their opinions by the defenders of Aristotle and Ptolemy on the one hand, and of the Copernican doctrine on the other.

No discussion could be undertaken without first enquiring into Aristotelian doctrines, which formed the basis of current theory. In a few words those amounted to a statement, that whereas things earthly are imperfect and full of change, things heavenly are eternal, unchangeable, and perfect. Salviati proves that this statement, in the spirit in which it was usual to accept it, was in reality untenable. The telescope showed him imperfections on the sun's surface, which was contrary to the belief that that body was unchangeable and free from blemish. He lays no great stress on the instance of comets (whose real nature Galileo never understood), but quotes the recent new stars (of 1572 and 1604) as instances of further change in the heavens. He thus prepares the way for a still wider departure from Aristotelian theory; he insists that the time has come to consider the nature of the world *de novo,* respectfully suggesting that Aristotle, had he the opportunities which the invention of the telescope afforded, would himself have been the first to realise the inadequacy of his dogmas on this subject.

Salviati proceeds to point out certain resemblances between the earth and moon and the more distant heavenly bodies. It is shown that the moon only shines in virtue of the sunshine falling on her. The idea that the earth might similarly appear luminous to any inhabitant—could one be imagined to exist—on the moon is less familiar, and less readily accepted. And yet the visibility of the moon during a total eclipse of the sun, and the appearance "of the old moon in the arms of the new" (as we now speak of it), are more probably due to reflected earth-light than to any other cause. This is discussed at some length, and the phenomena of Venus's phases (revealed by the telescope) are shown to be similar to those of the moon, and may be explained as due to the same cause. Venus then, like the moon, owes her brilliance to sunlight falling on her. The same probably applies to Mercury and Mars. The obvious inference seems to be that all of these heavenly bodies are not so unlike the earth as men had always been taught to believe. Points of resemblance there certainly are, and there may be many more, which the distance of the planets alone prevents us from discovering. Salviati refers to the common spherical form of earth, sun, moon, and planets, suggesting the existence of a common cause for that shape. The passage is striking enough to quote:

> Just as from the mutual and universal tendency of the parts of the earth to form a whole, it follows that they all meet together with equal inclination, and, that they may unite as closely as possible, they assume the spherical form, and so we ought to believe that the moon, the sun, and other mundane bodies are also of a round figure, if for no other reason than from a common instinct and natural concourse of their component parts; whence, if by accident any one should be violently separated from its whole, it is reasonable to believe that spontaneously and of its natural instinct it would return.

Having laid stress on the resemblance of earth and planets as a probable theory, Salviati proposes for them all a similar motion round the sun—one of the two main points of the new Copernican doctrines.

He shows how by this hypothesis the apparent paths of the planets can all be explained. And the simplicity of his explanation as contrasted with the Ptolemaic system appeals to common-sense, in a way which in itself almost carries conviction. A glance through a telescope turned towards Jupiter shows a family of small bodies circling round a great planet; here one could see on a small scale the very thing that Copernicus had described as going on in the case of planets and sun on a much larger scale, the sun being in the latter case the central body which corresponded with Jupiter in the other.

ON THE SECOND "DAY" the discussion passes on to the other chief point in the Copernican hypothesis, that the daily motion of the stars is only apparent, being due to a real daily rotation of the earth on its polar axis. Various objections are brought against this. The opponents of the earth's diurnal motion maintained that, if that motion were real, a stone dropped from the top of a tower would not fall at its foot. In the same way it was stated that a stone dropped from the masthead of a ship would fall near the stern, in consequence of the ship's velocity. But, strange as it seemed to Simplicio, it is nevertheless true that the stone falls at the foot of the mast, the ship's motion, provided it be uniform, having no power to disturb its fall. In a variety of forms the argument is brought forward, that if the earth be really rotating we on its surface ought to be sensible of the fact, either by our direct power of feeling, or else by irregularities in the motion of the things about us. There is a single reply to all such contentions, viz., all bodies on or near the earth's surface share the earth's motion, even the lower parts of the atmosphere being carried on by it, and, therefore, as all such things have a common motion, their relations to one another are just as if the motion did not exist. "Motion is so far motion (and as motion it operateth) by how far it hath relation to things which want motion; but in those things which all equally partake thereof it hath nothing to do, and is as if it never were." Salviati justifies this contention by a forcible illustration. A ship is out for many months on an ocean voyage, touching at various ports, and sailing now east, now west. Fix the attention on a single bale of cargo, packed tight in the hold; and, though the ship may have been tossed about in all directions by winds and waves, we are justified in saying that that particular bale of cargo has not moved during the long journey from port to port. Let the mate go down into the hold and disturb that bale from its place by one inch, and it has had in relation to the rest of the cargo a greater motion than it acquired during all the time that ship and cargo were voyaging together; the disturbances that were common to all had no visible effect inside the hold.

No objections to the hypothesis of the earth's rotation being found tenable, it is shown by Salviati how much more simple is the real motion proposed than the supposition that the universe revolves daily round a fixed earth. "To make the universe revolve," he says, "in order to maintain the immobility of the earth is as little reasonable as to require, in order to see Venice from the top of the Campanile, that the whole panorama should move round the spectator instead of his simply moving his head."

The primitive notion of the stars as fixed in a crystal sphere had been long overthrown. And, supposing accordingly that the stars were distinct and independent bodies, it was difficult to imagine laws controlling their motion about a fixed earth that should result in revolutions timed uniformly for all and at the same time of enormous rapidity. Salviati makes the improbable to be practically impossible by referring to the phenomenon now known as the "precession of the equinoxes," in virtue of which the direction of the earth's axis in space moves slowly, completing a revolution in about 26,000 years. The system of stellar motions that would be necessary to account for this would be inconceivably complex.

A GREAT PART of the third "Day" is devoted to the question of stellar parallax. In this lay one of the most serious difficulties of the Copernican theory. If it were true that the earth swept round the sun in a circular orbit some two hundred millions of miles in diameter, then it must follow that at one time of the year we should get an entirely different view of the arrangement of the stars from that obtained six months earlier or later when the earth was at the opposite point of its orbit. The nearer stars should in fact undergo displacements in their apparent positions relative to those more distant. The answer to this was that these displacements probably did take place, but were too minute to be detected. But this answer, though strictly true, implied that the distances of even the nearest stars were great beyond all comprehension; and this in turn implied that the visible size of the stars indicated a real size of inconceivable dimensions. The latter difficulty was reduced by Salviati's assertion that the visible size of a star was an optical illusion; the telescope showed the stars to be sharp points, in contrast to the planets which, though small to the eye, really did possess visible dimensions. But the former difficulty remained, and nearly two centuries passed before Bessel made the first rough measurement of a stellar parallax. His method was essentially that suggested in the dialogue by Salviati; though the results obtained indicated for the star 61 Cygni, a distance which would probably have astonished even Galileo himself.

With the difficulties of stellar parallax still uppermost in his mind, Simplicio looks at the utilitarian side of the question, and remarks:

> All this is very well, and it is not to be denied that the heavens may surpass in extent the capacity of our imaginations, nor that God might have created them a thousand times larger than they are. But we ought not to admit anything to be created in vain, or useless in the universe. Now we see this beautiful arrangement of the planets disposed round the earth at distances proportioned to the effects they are to produce upon us for our benefit. To what purpose, then, should such a vast vacancy be afterwards interposed between the orbit of Saturn and the starry spheres, containing not a single star, and altogether useless and unprofitable? To what end? And for whose use and advantage?

SALVIATI: Methinks we arrogate too much to ourselves, Simplicio, when we assume that the care of us alone is the adequate and sufficient work and limit beyond which the Divine wisdom and power do nothing and dispose of nothing. I feel confident that nothing is omitted by God's providence which concerns the government of human affairs; but that there may not be other things in the universe dependent on His supreme power, I cannot, with what power of reasoning I possess, bring myself to believe. So that when I am told of the uselessness of an immense space interposed between the orbits of the planets and the fixed stars, I reply that there is temerity in attempting by feeble reason to judge the works of God, and in calling vain and superfluous every part of the universe which is no use to us.

SAGREDO: Say rather that we have no means of knowing what is of use to us. I hold it to be one of the greatest pieces of arrogance and folly that can be in this world to say, because I know not what use Jupiter and Saturn are to me, that therefore these planets are superfluous. Nay more, that there are no such bodies in existence. To understand what effect is worked upon us by this or that heavenly body (since you will have it that all their uses must have a reference to us) it would be necessary to remove it for a while, and then the effect which I find no longer produced on me, I may say depended on that body. Besides, who will dare to say that the space (called too vast and useless) between Saturn and the fixed stars is void of other bodies belonging to the universe? Must it be so because we do not see them? Then, the four Medicean planets and the companions of Saturn came into the heavens when we began to see them, and not before! And by the same rule the innumerable host of fixed stars did not exist before men saw them. The nebulae, which the telescope shows us to be constellations of bright and beautiful stars, were, till the telescope was discovered, only white flakes! Oh, presumptuous! rather, oh rash ignorance of man.

Towards the end of the third "Day," reference is made to an annual rotation of the earth about an axis perpendicular to the plane of its motion. In ancient and mediaeval times a simple state of revolution of the earth round the sun would have implied a revolution in which the same side of the earth was always turned to the sun; the moon, according to this description, was said to revolve simply round the earth without any rotation of her own, which is not the way in which her motion would be described at the present day. Accordingly, in stating the earth to have a revolution round the sun combined with a rotation about an axis, Copernicus would have implied that that axis continuously changed its position in space so as to be always in the same direction relatively to the sun. To indicate that as a matter of fact the position of the axis with regard to the sun varied, but remained the same with regard to space, Copernicus had to combine with his two chief motions a third one of annual rotation.

This third rotation was therefore a complication only introduced by confusion in geometrical thought. That the actual state of things is quite simple is illustrated by Salviati by a reference to the motion of a ball floating in a basin of water. If the basin be held in the hand, the ball floating at or near the centre, and the experimenter turn round steadily on his feet, holding the basin in front of him, the ball remains in a position which is unaltered with reference to the walls and furniture of the room; although with reference to the man supporting the basin, it might be said to have spun once completely round. And so with regard to the annual rotation spoken of by Copernicus, Salviati says: "What other is the earth than a globe liberated in tenuous and yielding air? That which you think to be a revolution in itself, you will find to be a not moving at all, but a continuing to be altogether immovable in respect of all that is immovable."

It is here that he speaks so approvingly of the labours of his great English contemporary, William Gilbert of Colchester. He explains Gilbert's theory of the earth as a huge magnet, and develops it, mentioning incidentally some observations and experiments of his own on magnetic phenomena—notably on the increased power of magnets when suitably provided with armatures.

The attractive and repulsive properties of magnets reminded Simplicio that in considering the causes of natural phenomena, some effects are attributed to sympathy which is an agreement and mutual appetency between things having the same qualities, while other effects are due to antipathy when things naturally repel and abhor each other.

"And thus," cuts in the wag, Sagredo, "with these two words they are able to give a reason for a great number of effects which we see, not without admiration, to be produced in nature. But it strikes me that this mode of philosophising is not unlike the style in which one of my friends used to paint. On his canvas he would write with chalk: here a fountain with Diana and her nymphs; here some harriers; in this corner a huntsman with a stag's head; the rest may be a landscape of wood and mountain; and what remains to be done may be put in by the colour-man. Thus he flattered himself that he had painted the story of Actaeon, having contributed nothing towards it beyond the names!"

THE FOURTH "DAY" of the Dialogue is devoted entirely to an examination of the cause of the tides, and is a development and extension of his letter on the same subject to Cardinal Orsini, 1616. It is a singular circumstance that the argument, upon which Galileo mainly relied as furnishing a physical demonstration of the truth of the Copernican theory, rested on a misconception. The ebb and flow of the tides, he said, are a visible effect of the terrestrial double movement, since they are the combined result of (1) the earth's daily rotation, and (2) the inequality of the absolute velocities of the various parts of the earth's surface in its revolution round the sun. To this notion he attached capital importance, and he ridiculed Kepler's suggestion (which, however, was nearer the truth) that the attraction of the moon was in some way concerned in the phenomenon. That the influence of the moon was paramount had indeed been recognised in ancient times, but a Scientific explanation in detail was not to be expected until the law of universal gravitation had been fully realised.

This last part of the dialogue is therefore of little value, and may be passed over in considering the discussions of the first three "Days." The chief work of the dialogue was to establish the Copernican theory; which, first promulgated in the days when human vision was unaided, had been found by Galileo to be supported by all evidence that could be gathered by means of his new invention. The problem—if it may be still said to exist—takes a slightly different form at the present day. So far are we now from the pre-Copernican theory of a fixed earth, that we look upon no single object in the whole universe as fixed. The sun itself has its motion amongst the other visible stars; the present direction and rate of that motion are roughly known. Accordingly, the alternative which offered itself to the controversialists of Galileo's day, that either the sun or the earth was stationary, does not concern us; both of the bodies are moving. They move, however, in such a way that the motion of the sun is sensibly uniform, while the earth and the other planets can only be reasonably spoken of as travelling around him while he with his family of satellites is advancing through space.

With this understanding as to what is meant when we speak of the sun, though not stationary, being the centre of the planets' motions, it need scarcely be mentioned that the Copernican theory has acquired, since the days of Newton, an enormous mass of evidence, which in a work of the present type it would be out of place to discuss. But whereas Copernicus and Galileo made it a question practically for common-sense to choose between the simple and the geometrically complex, dynamical evidence has now made any alternative to the simple explanation not only difficult but altogether beyond comprehension; the main points of the Copernican doctrines are proved as absolutely as anything in Science can be proved.

This beautiful volume now so forgotten, of which we feel we have given an inadequate idea, is not simply a treatise on astronomical and physical Science—a powerful plea for Copernicanism in a country and at a time when all Science was "vehemently suspected"; it is a book worthy of Socrates—a book which ought to be studied by those who love free observation and experiment, free discussion and circulation of ideas—in a word, freedom of thought, as the first essential to the progress of Science and of our common humanity. In the words of Professor Playfair:

> One forms a very imperfect idea of Galileo from considering the discoveries and inventions only, numerous and splendid as they are, of which he was the author. It is by following his reasonings and by pursuing the train of his thoughts in his own elegant though somewhat diffuse exposition of them that we become acquainted with the fertility of his genius—with the sagacity, penetration, and comprehensiveness of his mind. The service which he rendered to real knowledge is to be estimated, not only from the truths which he discovered, but from the errors which he detected—not merely from the sound principles which he established, but from the pernicious idols which he overthrew. The *Dialogues on the Two Systems* are written with such singular felicity that one reads them at the present day, when the truths contained in them are known and admitted, with all the delight of novelty, and feels one's self carried back to the period when the telescope was first directed to the heavens, and when the earth's motion with its train of consequences was proved for the first time. Of all the writers who have lived in an age which was only emerging from ignorance and barbarism, Galileo has most entirely the tone of true philosophy, and is most free from the contamination of the time in taste, sentiment, and opinion.

11

PUBLICATION OF THE DIALOGUES

1632–1633

By the beginning of January 1632, the printing of the *Dialogues* was so far advanced, that on the 3rd Galileo was able to inform his friend Cesare Marsili at Bologna that the work would be ready in ten or twelve days. It did not, however, appear till February. On the 22nd of that month Galileo presented copies to the Grand Duke (his former pupil) to whom the work was dedicated, and to other members of the Medici family. Next day he sent thirty-two copies to Marsili; and had a number handsomely bound for friends and patrons in Rome, but they could not be despatched, owing to the continued prevalence of the plague. Indeed, it was not till May that two unbound copies reached the Eternal City. One of these came into the hands of Cardinal Francesco Barberini (the Pope's nephew), who lent it to Castelli. The latter writing to Galileo, 29th May, expressed his admiration for the work which surpassed all his expectations. Shortly afterwards, Conte Filippo Magalotti, Galileo's friend and, from his relationship to the Barberini family an influential personage, imported eight copies, and, at the author's request, presented one each to Cardinal Antonio Barberini, the Pope's brother; Niccolini, the Tuscan Ambassador; Father Riccardi, the Press Censor; Mgr. Serristori,

Counsellor of the Inquisition or Holy Office; and the Jesuit Father, Leon Santi.

While these copies were being eagerly read in Rome, and passed from hand to hand, the book had been circulating in all parts of Italy, in spite of the obstacles to communication caused by the plague. The applause with which it was received by all men of independent minds was tremendous; but in this paean of praise there was one solitary note of warning—a note which would probably have been unheeded even if it came in time. Paolo Aproino, having read a manuscript copy which Galileo had sent to Micanzio at Venice, begged his friend to write and advise the author to pause ere he printed a book containing such startling doctrines. Micanzio thought it better that Aproino should himself write, which he did on 13th March—a month too late. Aproino's advice was to send manuscript copies to the public libraries in the capitals of Europe, with permission for copies to be made by those who might wish to have them. This would prevent the dissemination of his revolutionary doctrines amongst the ignorant and ill-disposed who would only use them as a weapon for his destruction, while as to the enlightened and unprejudiced, no one would grudge the expense of a written copy of such a precious work.

Great as was the applause on the one side, so on the other side was the consternation which the book created among the followers of the old school of thought. The educated world of Italy was then divided into two hostile camps—that of Aristotle and Ptolemy on the one side; and that of Copernicus, Galileo, and Kepler on the other. In the first were to be found blind worship of authority, and unquestioning adherence to ancient doctrine; in the second, freedom of thought, research, recognition of demonstrated truths—in a word, Progress. As was to be expected, the first-named party was the most numerous and noisy, and it was reinforced by all those who opposed the innovators from interested motives. Foremost amongst these were the members of the Order of Jesus. They claimed for themselves the monopoly of instruction, and the first rank in the learned world, and were jealous of all intruders. Galileo was therefore in every way inconvenient to these people. Besides, had he not measured swords with distinguished members of the Order, as Fathers Scheiner and Grassi, and (unforgivable offence) had he not worsted them? And now his *Dialogues* appeared in which some old sores are re-opened; this revolutionary book must be suppressed at all costs, and with it its detested author.

FATHER RICCARDI, the censor, was the first to announce the coming battle. One day early in August he remarked to Conte Magalotti, "the Jesuits will now persecute Galileo with the utmost bitterness." It is important to establish this fact. Scheiner in a conversation with Torricelli would say little about the *Dialogues*, he found the digressions tedious (and no wonder, for some of them referred to himself), he did not wish to say much on the subject, but, he significantly added, "Galileo has treated me very badly." Then Scheiner himself, writing to Gassendi of Paris, 23rd February 1633, says:

> In these *Dialogues* the author has made null all my mathematical researches, and has laid violent hands on my 'Rosa Ursina,' on my discovery of the annual movement of the sun-spots, and of that of the sun himself. I am preparing to defend myself and the truth.

Before the date of his conversation with Magalotti just mentioned, it came to the ears of Riccardi, that some "ill-disposed persons" were trying to discover something in the book which could form the basis of an accusation against its author; they found something on the engraved title-page! The words "Dialogo di Galileo Galilei, Linceo, al Ser. Ferd. II Gran. Duca di Toscana" are printed on the field of a pavilion, with the five *palle* or balls, the armourial bearings of the Medici, and surmounted by the Grand Ducal crown. Below, on the shore of a sea stand three persons disputing—Aristotle, Ptolemy, and Copernicus, the two latter having their names printed on the edge of their mantles. At their feet is the device of three dolphins, surrounded by a narrow band, bearing a motto and the monogram GBL—the whole being the business sign of the printer, G.B. Landini. This title-page was impugned as not having been submitted for ecclesiastical approval, and particularly they expatiated, with more malice than wit, upon the meaning of the three dolphins' device; it reminded them so much of the three bees of the Papal arms! It was a great relief to Riccardi's mind when Magalotti pointed out that the same device appeared on nearly all the works which issued from the Landini press at Florence, where also this book was printed.

This shot, then, had not taken effect, and the "ill-disposed persons" had to find some other mode of attack. They now brought against the author the two-fold charge (1) that the preface was printed in different type from the rest of the book, which was true, but was simply a necessity of the printer; and (2) that some weighty arguments against the Copernican theory, which the Holy Father had brought forward in conversation with Galileo in 1624, were not in the printed book, although it was a condition that they should be. This charge was not true, for the "weighty arguments" were in reality only one, and this, as we have seen, was duly introduced at the end of the work, where it was appropriately put into the mouth of Simplicio, who gave it as an argument which he had from "a very eminent and learned personage" (p. 248 *ante*).

Foiled again, the assailants now fastened on the very natural circumstance that the "weighty argument" was placed in the mouth of Simplicio, the defender of Ptolemy. Knowing the Pope's weaknesses, vanity, arrogance, and ambition, they made him believe that by Simplicio—the simpleton—no other was intended than Urban VIII himself! One would have thought this impossible with this shrewd old man (seeing the friendly relations which up to this had subsisted between him and Galileo), but it is beyond all doubt that it was so, and it put his Holiness in a terrible rage. While in this condition they easily persuaded him that the Copernican doctrine ran counter to the dogmas of the Catholic faith, and after this it was not difficult for them to show that the *Dialogues*, which were a defence of that doctrine under a flimsy veil, would do incalculable injury to the Church. Having thus worked on the Pope's fears, they easily persuaded him that in this work Galileo had again, though under concealment, dared to interpret the Scriptures; he was therefore rebellious; and he was further deceitful in that he obtained the Imprimatur by cunning devices.

Although the safety of the Church and the vindication of its decrees were the ostensible reasons for the subsequent proceedings against Galileo, it would not be far from the truth to say that revenge for an assumed personal insult was the primary and determining factor. Without this personal motive the storm may have blown over as other and equally threatening ones had previously done. Urban VIII and many of the high dignitaries of the Church were, if anything, Copernicans (they certainly were not Ptolemaics, and hardly even peripatetics), while others were indifferent and cared little one way or the other. As regards the Pope himself, we have seen his letter of 20th April 1613, to Galileo, praising so highly the book on sun-spots—the very work on which the prohibition of 1616 was based. We have also seen how delighted he was with "Il Saggiatore" and the "Reply to Ingoli," in both of which the Copernican theory is defended; and, finally, we have Urban's own statements (1) that the Copernican doctrine is not heretical, but only rash, and (2) that if it rested with him, the decree of 1616 would never have been issued. All this seems to show that, if the question of a personal insult had not arisen, the *Dialogues* might have weathered the storm, or, at the worst, been put on the Index, as was the book of Copernicus in 1616, "until corrected."

While these things were passing secretly at Rome, Galileo in Florence gave himself up to unmixed delight at the great success of his book. His friends, such as Castelli, Cavalieri, Micanzio, Campanella, and others, expressed in letters their unbounded admiration. Thus, Micanzio, writing on 3rd July, says:

> I had hardly time to devour your book when it was taken from me and lent from one to another. Today, no sooner do I get it back by main force, than I am obliged to send it to the

> Commissary Antonini at Verona, one of our cleverest men, and one who admires you above all the literati of the age.

Not one of all his friends who praised the book so highly had any foreboding that it was soon to bring its grey-headed author before the bar of the Inquisition, and least of all Galileo himself. He, of course, expected the usual opposition from his "Scientific" opponents, and was prepared to meet it; but he considered himself secure from anything like conflict with the ecclesiastical authorities.

ONE DAY, about the middle of August, the first thunder-clap broke over Galileo. His publisher Landini received instructions, though for the time only provisional, forbidding the further sale of the *Dialogues*. The second clap, which followed in a few days, was the news of the appointment of a special commission at Rome by order of the Pope, and under the presidency of his nephew, Cardinal F. Barberini, to examine the book and report. It was composed of Jesuits, Dominicans, and Theatines, "not one of whom had any knowledge of mathematics, or familiarity with abstruse speculations," all endeavours of Niccolini and others to get friends of Galileo, such as Castelli and Campanella, put on the board being vetoed by the Pope. Galileo now appealed to his Sovereign for protection, and, on 24th August, the chief Secretary Cioli wrote to Niccolini that the Grand Duke was greatly surprised to learn that a book which had been laid before the supreme authorities at Rome by the author in person; had been carefully read there again and again, as well as afterwards at Florence; had, at the author's request, been altered by these authorities "as seemed good to them"; and finally had received the Imprimatur for Rome and Florence, should now, after two years, be considered suspicious, and be prohibited. His Highness was of opinion that this opposition must be directed against the person of the author rather than against his book, so often and so carefully read and revised by the proper authorities. In order, then, to inform himself of the merits or demerits of his servant, his Highness desires that that which is granted in all disputes and before all tribunals should be permitted to the accused, viz., to defend himself against his accusers. The Grand Duke, therefore, requests that the accusations may be sent to Florence, so that the author, who stands firmly on his innocence, may see and answer them.

On the same day on which this despatch left Florence a mandate was issued from Rome, confirming the provisional prohibition of the *Dialogues*, and ordering Landini to send all copies in stock to Rome. He replied that the edition had been sold out, and, consequently, he had no copies to send.

On receipt of the Grand Duke's orders, Niccolini hastened to carry them out, but met with more opposition than either he or his master had expected. The Pope received him in such a way that he thought the world must be going to pieces, and at the first mention of Galileo's name the Holy Father interrupted him bluntly: "Your Galileo has dared to meddle with things that he should have left alone—in fact, with the most important and dangerous subjects that can be stirred up in these days." On the Ambassador remarking that the book was published with the approbation of the Church, the Pope angrily replied that both Galileo and his secretary, Ciampoli, had deceived him on that matter. The censor, Riccardi, had also deceived him, "but," he added, "the latter had been himself deceived, for he had been induced by fair speeches to approve the book, and by more fair speeches to allow it to be printed in Florence." Thinking to save the book from condemnation, Niccolini hinted that it was dedicated to the Grand Duke. "What of that?" was the reply, "I have prohibited books which bear my own name on the title-page." The Pope added, "In charging a special commission to examine the book, instead of handing the affair at once to the Inquisition, I have followed a course best for Galileo's interests—he who did not fear to make game of me." Niccolini then begged that the accused may know the charges against him and have an opportunity of justifying himself, to which Urban curtly answered: "Galileo knows well enough in what way he has transgressed. In these matters of the Holy Office nothing is ever done but to pronounce judgment, and then summon to recant."

Two letters from Magalotti, who was usually well informed, arrived at the same time as this despatch, both dated 4th September, one to Mario Guiducci, the other to Galileo. Magalotti's news was on the whole reassuring. From the opinions of persons present at the sittings of the commission, he thought he could say that matters would not go so far as condemnation of the Copernican doctrine "by supreme authority." He thought with Riccardi that they would not entirely prohibit the *Dialogues*, but only "correct" them, so as to sustain the decree of 1616. He advised (as Niccolini also had done) the utmost patience and circumspection, and to confer with Cardinal F. Barberini rather than with the Pope himself, "for reasons which it is not necessary to detail."

The special commission, after a month's session, submitted a long report to the Pope. The document begins with a statement of the course of the negotiations for the printing of the *Dialogues*, and then come three indictments against the author:

1. Galileo had transgressed orders in deviating from the hypothetical standpoint, by maintaining decidedly that the earth moves and that the sun is stationary.
2. He has erroneously ascribed the phenomena of the tides to the stability of the sun and the motion of the earth, which are not true.
3. He has been deceitfully silent about the command laid upon him in 1616, viz., to relinquish altogether the opinion that the sun is the centre of the world and immovable and that the earth moves, nor henceforth to hold, teach, or defend it in any way whatsoever, verbally or in writing.

Then follows the remark: "It now remains to be considered what proceedings are to be taken against the person of the author, and against his printed book." The rest of the document is taken up with an elaboration of the charges against Galileo, and a fuller account of the negotiations for the Imprimatur. These need not detain us; but in a final clause, Galileo is specifically charged with having transgressed the order of the Holy Office to relinquish, etc., as per charge (3) above. This, then, was his chief offence; the others "could be corrected if the book was thought to be of any value"; but to deliberately and deceitfully transgress commands, that, evidently, could not be "corrected" or condoned.

Here for the first time the minute of 26th February 1616 is mentioned, as to which Niccolini has something of importance in his despatch of 11th September to Cioli. Reporting an interview with the Master of the Sacred Palace, he says that Riccardi, after advising patience and caution, repeated the old stock complaints, that the *Dialogues* imperilled the faith, that the author did not confine himself strictly to mathematics, but brought under discussion religion and the Scriptures [which is not true], and that the Papal orders as to the preface, end, and hypothetical treatment had not been complied with —an extraordinary assertion for one who had himself certified that they had been complied with. He then confided to the Ambassador as a profound secret "that it had been discovered in the books of the Holy Office that sixteen years ago (it having been heard that Galileo entertained that opinion, and disseminated it in Florence) he was summoned to Rome, and forbidden by Cardinal Bellarmine, in the name of the Pope and the Holy Office, to hold that opinion. And this alone is enough to ruin him entirely." Evidently the paper here hinted at is the minute of 26th February 1616.

It will, therefore, be desirable to pause for a moment, and read again our remarks on the first Inquisition, for our account so far of these new proceedings brings out a point there discussed, which we wish to emphasise, viz. that the minute is not true to the spirit of the proceedings. We have seen that the Pope knew all about the process of 1616—in fact, as Cardinal Maffeo Barberini, was an unwilling party to it. If Galileo was then commanded to relinquish altogether his opinion, and henceforth not to hold, teach, or defend it in any way whatsoever, verbally or in writing, why did Urban VIII discuss this same opinion with him in 1624? Why did he openly approve "Il Saggiatore" and the

"Reply to Ingoli"? And why in 1630, when Galileo took his *Dialogues* to Rome, was he not met with a ***non-possunius,*** and referred back at once to this rigid prohibition? Clearly, because this rigid prohibition had for him no existence—had no right to exist, for he knew it was not intended that Galileo should be forbidden to hold, teach, or defend his opinion in any way whatsoever, verbally or in writing, except in case he refused to be bound by Cardinal Bellarmine's simple admonition—a contingency which did not arise. The minute, therefore, is not true to the spirit of the proceedings, and should properly be treated as non-existent. It is, moreover, of no value in law in that it bears no signatures. To base, then, a charge against Galileo on such a worthless paper convicts his persecutors of *ultra vires,* and shows their determination to punish him at any price, even at the cost of a judicial crime.

NOW LET us return to our narrative. A few days after Niccolini's interview with Riccardi, on the 15th September, the Pope sent word to the Tuscan Ambassador that Galileo's affairs would be handed over to the Inquisition. At the same time the strictest secrecy as to this step was enjoined on both the Grand Duke and Niccolini, with a hint that otherwise they would be proceeded against, according to the statutes of the Holy Office. Niccolini, astounded by this intimation, hastened to the Pope to avert, if possible, the danger of a trial, but his pleadings were in vain. Urban was good enough to say that Galileo was still his friend, but his opinions had been condemned sixteen years before, and were in the highest degree pernicious to the Church. On 23rd September the following order was issued:

> His Holiness charges the Inquisitor at Florence to inform Galileo, in the name of the Holy Office, that he is to appear in the course of the month of October, in Rome, before the Commissary-General. He must obtain a promise from Galileo to

> obey this order, which the Inquisitor is to deliver in the presence of a notary and witnesses, but in such a way that Galileo shall know nothing of their presence, so that if he refuse to obey they may bear witness to his contumacy.

This order, which was delivered to Galileo on 1st October, and with which he consented in writing to comply, fairly overwhelmed him, for, from the secrecy maintained in Rome, he was wholly unprepared for any such measure. Scarcely recovered from a complaint in the eyes which had lasted several months, suffering otherwise in health, and at an advanced age, he was now to go to Rome in the midst of the plague (which had broken out afresh) to appear before the terrible Inquisition. No wonder that he was dismayed, and that in spite of his promise "willingly to obey the order in the course of this month, October," he made every effort to evade it. His deep depression is evident from a long and pitiable letter, of 13th October, to Cardinal Francesco Barberini (the Pope's nephew), sent through the hands of Niccolini:

> That my *Dialogues* recently published should find adversaries was not to be doubted, as your Eminence no doubt foresaw. It is in general the lot of all opinions which in one way or another run counter to the accepted doctrines. From the reception of my other works I expected as much, but what I did not expect was that the hate of one or two of my enemies (furious at seeing the lustre of their works tarnished by mine) would be able so to influence my superiors as to make them believe that my works are unworthy of the light of day, and should be stifled. The prohibition of the printing and sale of my *Dialogues* has been a cruel blow to me, but I am consoled by the knowledge of the extreme purity of my conscience, and by the feeling that I shall have no difficulty in justifying my intentions.
>
> I will not conceal from your Eminence that the injunction to

present myself without delay before the tribunal of the Holy Office has afflicted me profoundly. It is impossible to think without bitterness that the fruits of my labours and studies for so many years (which gave to my name in the Scientific world a certain *éclat*) should now be branded as criminal. All this depresses me to such an extent as to make me curse the time I have devoted to these labours—yes, I regret having given to the world so much of my results. I feel even the desire to suppress, to destroy for ever, to commit to the flames, what remains in my hands. Thus I should satisfy the burning hate of my enemies. These are some of the thoughts which afflict me, and increase the burden of my seventy years; they aggravate my numerous physical sufferings, and cause me persistent insomnia. When to these is added a journey, rendered more painful and dangerous by sundry causes, I am almost certain that I shall not reach the end alive. The desire to live, common to all men, makes me implore the intercession of your Eminence, encouraged thereto by the kindness of heart which distinguishes you, and of which I as well as others have been the recipient. I beg you, then, to represent to the Holy Father my present pitiable situation. [...]

Whether it be necessary to receive my justification in writing, or by *viva voce,* I would point out that there are here in Florence the Inquisitor, the Archbishop, and other learned functionaries of the Church, who would be able, it seems to me, to decide graver causes than mine, and before whom I am ready to appear. It is hardly likely that in a book, which has been carefully examined by the censors, with full power to omit, to add here, to correct there, there should still remain errors so grave that their correction, or the punishment due to them, should be beyond the power of the local authorities.

Before taking action the Ambassador consulted Castelli (whom the Grand Duke had appointed as his counsel in this affair) as to the advisability of delivering Galileo's letter to the Cardinal. It was decided to do so, and Niccolini reported that it was received in a very friendly way, with assurances of his Eminence's kindly disposition towards the writer. The letter was discussed by the Holy Office on 11th November, in presence of the Pope; but his Holiness would not grant the prayer, and ordered that the writer must be compelled to come to Rome. Niccolini was unwearied in trying to get the affair settled in Florence, but to no purpose. The Pope had seen Galileo's letter of 13th October, but the journey to Rome could not be dispensed with. "Your Holiness," exclaimed Niccolini, "incurs the danger, considering Galileo's age, of his being tried neither in Rome nor in Florence, for I solemnly assure your Holiness that he may die on the way under all these difficulties and anxieties." "He can come very slowly," replied the Pope, "in a litter, with every comfort, but he really must be tried here in person. And may God forgive him for having been so deluded as to involve himself once more in these difficulties, after having been extricated by me from his first difficulties in 1616, when I was Cardinal." Much discomfited and with profound sorrow, Niccolini communicated this decision to Galileo in a letter on 13th November, and in a despatch to Cioli of the same date.

A few days later, 19th November, Galileo was summoned before the Inquisitor at Florence, in accordance with the Papal orders, and was charged to comply with the mandate to go to Rome in the presence of a notary and two witnesses—a respite of one month being allowed. The appointed time had nearly arrived and no preparations had been made for starting. The Inquisitor sent his vicar to see Galileo and reported the result in a letter to Rome on 18th December:

> My vicar found Galileo in bed; he was quite ready to set out, but in these times he had no heart for it; besides, just now, owing to a sudden attack of illness he was not in a condition to travel. He has sent me the enclosed medical certificate, so that I have not failed to do my duty.

The certificate, dated 17th December, gives an idea of the physical sufferings of this much-tried man, and is signed by three doctors:

> We, the undersigned physicians, certify that we have examined Signor Galileo Galilei and find that his pulse intermits every three or four beats, from which we conclude that his vital powers are affected, and at his great age much weakened. To the above are to be ascribed frequent attacks of giddiness, hypochondriacal melancholy, weakness of the stomach, sleeplessness, and flying pains about the body, to which others also can testify. We have also observed a serious hernia with rupture of the peritoneum. All these symptoms are worthy of notice, as under the least aggravation they might become dangerous to life.

Little importance seems to have been attached to this certificate at Rome. Niccolini (26th December) fears that the ecclesiastical authorities at Florence will be ordered to take extreme measures. Castelli (25th December) urges his old and revered master to set out. But in this, as in all his letters of this period, he shows that he had no idea of the real nature of the proceedings going on. Knowing of no crime committed by his master against the Holy Office, he urged him to set out, because he had the idea that Galileo's persecutors desired nothing better than that he should refuse, in order that they may decry him as an obstinate rebel against the Holy Church.

On 30th December the fears of Niccolini were realised. On that day a Papal order was issued to the Inquisitor at Florence, stating that neither his Holiness, nor the Congregation of the Holy Office, could or would tolerate further evasions. It must, therefore, be proved that Galileo's state was really such that he could not come to Rome without danger to his life. A commissioner with a physician would be sent to Florence, who would see Galileo and make a true and trustworthy report on his condition, and if he were in a state to travel the commissioner must bring him a prisoner in chains. If out of consideration for his health, or other danger to life, his coming must be postponed, then as soon as he had recovered, or the danger was over, he was to be brought a prisoner in irons. The commissioner and physician would travel at Galileo's expense, because he had not obeyed the command to appear when his condition would have permitted it.

To avert these extreme measures the Grand Duke caused Galileo to be informed (11th January 1633) that it was at last necessary to obey the orders of the supreme authorities at Rome, and in order that he might perform the journey more comfortably Grand-ducal litters and a trustworthy guide would be placed at his disposal; he would also be lodged in Rome in the house of the Grand Duke's Ambassador. The pitiful impotence of an Italian ruler of that day in face of the Roman Church is painfully obvious in this decision. The Sovereign does not dare to protect his subject—more, his old and respected tutor, and the greatest philosopher of whom Italy could boast—but gives him up to the dreaded Inquisition, as if he were an alien malefactor. The Venetian republic was the only State in Italy that would have asserted its independence (as it had often done before), and would have refused to hand over one of its officials to the Roman power. Indeed, when these proceedings began, Francesco Morosini of Venice offered to reinstate Galileo in his old chair at Padua on any conditions that he chose to make, and to print his *Dialogues* in Venice. Galileo was now suffering a bitter penalty for the mistake of 1610 in deserting Padua.

ON 20TH JANUARY, this man of woes set out on his terrifying journey in a Grand-ducal litter. How different the circumstances from those of the same journey eight years before! Prince Cesi, his entertainer at Acquasparta, was dead; his own health, never good for many years past, was now a chronic cause of suffering; and his eyesight had begun to fail. Not only was the time of year unfavourable (January–February being the season when the biting tramontana is most frequent), but the country through which he had to pass was bleak and inhospitable, and its inhabitants, always wild as the winds that howl across its wastes, were now made more wild and desperate by the ravages of the plague. At the frontier post, Ponte Centino, he was obliged to halt twenty days on account of quarantine. From each halting place the poor old man had written to his daughter, the one soul on earth to whom he knew he could turn for sympathy and consolation, and she, sweet consolatrix, what can she say in reply? After expressing her grief at his being detained so long in a wretched habitation, deprived of every comfort, she entreats him "to keep up his spirits, and to put his whole trust in God, who never forsakes those who trust in Him." What else could she say? Although at the moment she knew that some of God's ministers on earth were intent on his persecution for the greater honour and glory of His kingdom.

On the afternoon of 13th February Galileo arrived in Rome and was warmly received at the Tuscan Embassy. On the next day Niccolini informed the Grand Duke of his arrival, and that next morning he would introduce him to Cardinal F. Barberini, and beg that he be permitted to remain at the Embassy instead of being locked up in the prison of the Holy Office. This favour was at once granted provisionally, and afterwards officially confirmed, with injunctions to keep indoors, and to see no one until further orders. Beyond this, and to Galileo's surprise, no official notice was taken of his presence for some time. Writing to Cioli on 19th February, he says:

> As to the situation of my affairs I can tell you nothing. However, to judge by what has passed, it appears to me, as well as to the Ambassador and his staff, that the storm which menaces me is a little calmed—at least in appearance, so that I do not give way to discouragement, as if shipwreck were inevitable, and all hope of reaching port were gone—the more so as following the instructions of my master, Ariosto:
>
> *I make sail with modesty*
> *Amidst the raging billows.*

> I keep to the house, not thinking it proper to go out as if I wished to show myself. Up to the present no official steps have been taken. One of the members of the Congregation has called on me twice, conversing in the most agreeable manner, and giving me an opportunity to explain myself, and to show my submission, always sincere, to the Church—to which, as far as I could see, he listened with satisfaction. If his visits, as may be supposed, were made with the knowledge or even by order of the Congregation, I may consider them as the beginning of a milder treatment, far removed from the cords, chains, and dungeons, with which I was menaced. I find another consolation in the kindly sentiments towards me expressed by many influential personages. As it appears to me easier to confirm the latter in their good opinions than to convert those who are unfavourable, I think, with the Ambassador, that letters from our august master to Cardinals Scaglia and Bentivoglio would be very useful, and if you agree with us I beg that you will obtain me this favour.

The member of the Congregation here referred to was Mgr. Serristori. Niccolini thought that the object of his visits was to discover the present sentiments and defensive arguments of the dreaded dialectician, and was inclined to think that they boded well.

"I think," he writes, "I have succeeded somewhat in cheering up the good old man by what I have told him of the steps being taken in his favour. But he constantly expresses his wonder at all this persecution."

The same cheerful confidence is expressed in Galileo's letter of 25th February, to Bocchineri:

> The Ambassador thinks he perceives each day a diminution in the irritation against me, and so does Father Castelli, who is for me a zealous and indefatigable advocate. We learn, in short, that the many and serious accusations against me are reduced to one,

and that the others have been allowed to drop. Of that one I shall be able to clear myself without much trouble, when the grounds of my defence have been heard. Little by little I am bringing these to the ears of some of the higher officials, who can neither refuse absolutely to listen to my explanations, nor to leave them entirely without reply. So we conclude that in the end a favourable issue may be hoped for.

I keep strictly to the house which appears to me and my friends to be the correct thing. The Cardinal Barberini has given me the same counsel, not *ex officio* but, as he says, as a friend. As I have already told you, not a word official has yet come from the tribunal. One of the counsellors, my friend and protector for many years, has visited me a couple of times and furnished me an occasion to explain myself freely on several points, and to show him some papers drawn up by me for this case, with which he declared himself satisfied. We suppose, not without reason, that these visits have been made with the knowledge, and, perhaps, even by order of the superior authorities, with the object, perhaps, of getting some general information. If so, they could not have adopted a better course in my interests.

The deprivation of exercise for the last forty days begins to be prejudicial to me, for as you know I find exercise necessary for my health. My digestion especially is troubled; viscous matters accumulate; and for the last three days painful twitchings in the limbs have prevented my sleeping. I hope that a severe dietetic regimen will effect a cure. Some days ago I told you how useful would be letters from his Highness to Cardinals Bentivoglio and Scaglia, who, as I am privately informed, are well disposed towards me. If we find in the Congregation one or two members ready and resolute to defend innocence and the truth, we may hope that their voices will suffice to impose silence on those inimically inclined. Therefore I beg you to procure these letters through the medium of Signor Cioli.

P.S.— Please communicate this letter to my daughters and Vincenzio.

Niccolini's despatch to Cioli of two days later (27th February) explains the nature of this chief accusation:

> Although I am unable to say precisely what stage Galileo's affairs have reached, or what may happen next, as far as I can learn the main difficulty consists in this—these gentlemen maintain that in 1616 he was commanded neither to discuss the question of the earth's motion, nor to converse about it. He says, on the contrary, that these were not the terms of the injunction, which were that that doctrine was not to be held or defended. He considers that he has the means of justifying himself, insomuch as it does not at all appear from his book that he does hold or defend the doctrine; or that he regards it as a settled question, as he merely adduces the reasons *hinc inde.* The other charges appear to be of less importance and easier to get over.

On the same day the Ambassador in a long audience officially announced Galileo's arrival, and expressed the hope that his Holiness would now be convinced of his reverence for things spiritual, especially in reference to the matter in hand. The Pope replied that he had shown Galileo a special and unusual favour in allowing him to stay at the Ambassador's house instead of remitting him to prison. Niccolini suitably acknowledged the great favour, and then went on to urge that in consideration of Galileo's age and bad health the trial may be hastened. Urban replied that the proceedings of the Holy Office were usually tedious, and he really did not know whether so speedy a termination could be looked for. They were still engaged with the preliminaries.

~~CHRIST PRESENTED TO THE PEOPLE~~
GALILEO PRESENTED TO THE INQUISITION

Pope Urban VIII said unto them, "Shall I excommunicate your Teacher?"

As to the nature of these preliminaries nothing certain could be learnt, but Galileo continued to hope for the best. In a letter to Geri Bocchineri (5th March), after acknowledging receipt of the letters for Cardinals Bentivoglio and Scaglia which had, as he thought, a good effect, he goes on to say:

> As to my affair, it goes on silently as from the first day. If one can judge by rare signs, the accusations have lost much of their gravity, and already some have been entirely dropped by reason of their evident insignificance, which is a good presage for those that remain.

A fortnight later, on 19th March, he wrote in a similar strain to Cioli:

> The same silence continues to be observed in my affair, and nothing more can be ascertained than what the Ambassador has picked up here and there, and which is sufficiently vague. My indefatigable defender, Benedetto Castelli, has also been secretly informed, but in the same general terms, that the proceedings have taken a slightly more favourable turn, thanks to the letters of his Highness. Therefore, as the Ambassador will tell you, the same intervention with the other Cardinals, members of the Congregation, would be of great utility, the two to whom letters have already been addressed having expressed themselves in this sense. I pray you then to obtain for me from his Highness this additional favour.

In spite of the gravity of the situation, it would seem that Galileo was careful to write to his daughter in such a strain as to calm her anxiety throughout this weary time. Therefore, on 13th March, she wrote back:

> As matters are going on so favourably I will not mind though your return be delayed, for indeed my being disappointed is a small thing, if staying where you are redounds to your reputation and advantage; and what makes me still more easy is to hear how honourably you are treated by those excellent gentlemen, and in particular by her Excellency the Ambassadress. I am well now because my mind is at rest. Nevertheless I do not cease praying for you.

On the same day (13th March) on which Maria Celeste was thus hopefully writing, Niccolini had an audience of the Pope, in which he was informed that it would be necessary to summon Galileo to the Holy Office on the eve of the trial. It was the usage and could not be departed from. Again, the Ambassador pleaded the accused's bad health, great age, and the willingness he had shown to submit to any penalties; but Urban replied it would not do to order otherwise. He regretted that Galileo "who had been his friend, with whom he had often held confidential intercourse, and eaten at the same table," should be subjected to these annoyances, but it was in the interests of religion.

Notwithstanding this intimation, days and weeks passed, and yet Galileo was not summoned before the Inquisition. All this time, as we see from his letters, he was entertaining confident hopes of some favourable issue; but, as a matter of fact, there were no grounds for this belief. Neither he nor his indefatigable friends, Castelli and Niccolini, could learn anything definite. The members of the Congregation, who alone could have given information, kept the secrets of the Holy Office very closely, as indeed they were bound to do under the severest penalties to themselves. Thus the month of March passed by; April was come, and with it the storm which had been so long threatening.

On the 7th, Niccolini went to Cardinal Barberini by invitation, and was informed, on behalf of the Pope and the Congregation, that Galileo must appear before the Holy Office, and, as it was not known whether the case could be settled at a single sitting, it might be necessary to detain him. Once more the Ambassador urged consideration for his age and health (he had again been ill and confined to bed), and begged the Cardinal to consider whether it would not be possible for him to return every evening to sleep at the Embassy. To these appeals his Eminence replied that such a permission was not to be expected. He promised, however, that every comfort would be afforded him in the buildings of the Holy Office; that he would neither, as was customary with accused persons, be treated as a prisoner, nor be placed in a cell; but he would have good rooms, and, perhaps, his doors would not even be locked. Niccolini reported this notification to Cioli on 9th April, adding the following interesting details:

> This morning I also had a conversation with his Holiness, who again gave vent to his displeasure that Galileo should have discussed this subject which appears to him to be very serious, and of great moment to religion. Signor Galileo thinks, nevertheless, that he can defend his statements on good grounds; but I have warned him to refrain from doing so, in order not to prolong the proceedings, and to submit to what shall be prescribed to him to believe respecting the motion of the earth. He has fallen into the deepest dejection, and since yesterday has sunk so low that I am in great concern for his life. I shall beg that a servant may be allowed him, and as much comfort as the place will admit of. Meanwhile, we are all doing our best to console him and to help him through our recommendations to the most friendly disposed members of the Congregation; for truly he deserves every possible kindness that can be shown him. I

cannot describe to you the grief of the whole house, for everyone here loves him exceedingly.

From this, then, we learn that up to 8th April Galileo was still intending to defend himself and his opinions, and that it was only on the earnest entreaty of the Ambassador that he gave up all idea of opposition, and resolved upon entire submission.

12

GALILEO AND THE INQUISITION

1632–1633

On 12th April 1633, Galileo appeared for his first examination before the Commissary-General of the Inquisition, Father Firenzuola, and the Procurator-Fiscal, Father Sincero. In all his answers he is actuated by one idea—that of shortening the proceedings and averting severe measures by submissive acquiescence. According to the rules of the Holy Office, an oath is administered to the accused that he will speak the truth, and he is then asked whether he knows or conjectures the reason of his citation. Galileo replied that he supposed he had been summoned to give an account of his last book. After being asked if he acknowledged the work shown him, "Dialogo di Galileo Galilei, Linceo," as his, and his reply in the affirmative, the examiners led him back to his visit to Rome in 1616, and to what then happened in the matter of the prohibition. His answers show that he only knew of Cardinal Bellarmine's admonition, as recorded in his certificate of 26th May 1616. That a "command" in more stringent terms was issued to him (as the Inquisitor asserts) he is not aware, but, true to his resolve of submissiveness, he says again and again: "It may be so, but I do not remember it." He is then told that this "command" was to the effect that he must not hold, defend, or teach, in any way whatever, verbally or in writing, the doctrines of

Copernicus. Galileo, who hears for the first time the exact terms of this further injunction "not to teach in any way whatever, verbally or in writing," is amazed, but is still submissive. "It may be," he answers, "that this injunction was also there, but I do not remember it." He then appeals again to Cardinal Bellarmine's certificate, "in which," he says, "there is no mention of this further injunction which has just been made known to me."

After trying by every artifice to get from him an admission that this further injunction was laid upon him, and after his declaring for the fifth time that he did not remember any "command" beyond the admonition of Cardinal Bellarmine, the Inquisitor asked whether he had received any permission to print his *Dialogues*, and if in the negotiations he had mentioned the "command" aforesaid. He replied: "I did not say anything about it when I asked for the Imprimatur. I did not think it necessary to say anything, because I did not consider that in writing the book I was acting contrary to, far less disobeying, the command 'not to hold or defend' the aforesaid opinions. I have neither maintained nor defended the opinion that the earth moves and that the sun is stationary, but have demonstrated the opposite, and shown that the arguments of Copernicus are weak and inconclusive." With this answer the first sitting of this memorable trial closed. Silence on matters connected with the proceedings having been imposed on oath, Galileo was led to his apartments in the private quarters of the Procurator-Fiscal, where he received kind and considerate treatment. Writing to Bocchineri on 16th April, he says:

> Contrary to custom, three large and comfortable rooms have been assigned to me, with permission to walk about in the spacious corridors. My health is good, for which, next to God, I have to thank the great care of the Ambassador and his wife, who have a watchful eye for all comforts—far more than I require.

A servant was allowed to remain with him, Niccolini was permitted to send in his meals, and no obstacle was opposed to his free correspondence with the kind Ambassador.

In their treatment of Galileo personally (unprecedented for its considerateness), throughout the whole of these proceedings, the Inquisition did homage to his genius. Since the establishment in 1215 of this dreadful tribunal, no prisoner had ever been treated with such leniency; for princes, prelates, and noblemen, all had been consigned to the secret dungeons from the very commencement of their trial. The Pope himself in an interview with Niccolini (reported in his despatch to Cioli, 27th February 1633) said, apropos of this point, that Galileo was treated as a privileged person, and reminded the Ambassador that in the recent case of a gentleman of the princely house of Gonzaga far different measures had been taken, that he had been not only carried to Rome by officers of the Inquisition, but was kept closely imprisoned for a long time before judgment was pronounced.

On 15th April, three days after the first examination, three counsellors of the Holy Office, Oregius, Inchofer, and Pasqualigus, delivered their opinions on the case. Oregius declared that in the *Dialogues* the doctrine that the earth moves and that the sun is stationary is held and defended. Inchofer declared that Galileo not only taught and defended that doctrine, but was himself suspiciously inclined to it, and even held it to this day. Pasqualigus was of opinion that by the publication of his *Dialogues* Galileo had infringed the order of 1616 in respect to teaching and defending, and it was very suspicious that he held the prohibited doctrine.

AFTER THE FIRST hearing of his case on 12th April, weeks passed and no further open step was taken in the trial; meanwhile, the prolonged deprivation of exercise in the open air which had been so essential to his health, combined with mental agitation, threw the old man on a sick bed. Writing on 23rd April to Geri Bocchineri, he says:

I am writing in bed to which I have been confined for sixteen hours with severe pains in my loins, which according to my experience will last as much longer. A little while ago I had a visit from the Commissary and the Fiscal who conduct the enquiry. They have promised and intimated it as their settled intention to terminate the case as soon as I am able to get up again, encouraging me repeatedly to keep up my spirits. I place more confidence in these promises than in the hopes held out to me before, which as experience has shown were founded rather upon surmises than real knowledge. I have always hoped that my innocence and uprightness would be brought to light, and I now hope it more than ever.

Please send this on to my daughters and Vincenzio, as usual.

The second examination was fixed for 28th April, and the course it now took had up to recent years puzzled all students of this famous trial. While at the close of his first examination we have seen Galileo deny having defended the Copernican doctrine, and assert that he had done just the opposite, now at this second hearing and almost without waiting for the Inquisitor's questions, he makes a declaration which, roundabout though it is, contains a penitent confession that he had defended those doctrines. The cause of this change is explained by a letter from one of Galileo's judges, the Commissary-General Firenzuola (who was then with the Pope in the Castle of Gandolfo), to Cardinal Francesco Barberini, and dated 28th April. This interesting letter (slightly condensed) runs as follows:

In compliance with the commands of his Holiness I yesterday informed the Congregation of the state of Galileo's case. Their Eminences approved of what has been done thus far, and took into consideration various difficulties with regard to the manner of pursuing the case, and of bringing it to an end. More especially

> as Galileo has in his examination denied what is plainly evident in his book; and in consequence of this denial there would result the necessity for greater rigour of procedure, and less regard to the other considerations belonging to the business. Finally, I suggested that the Congregation should grant me permission to treat extrajudicially with Galileo in order to render him sensible of his error, and bring him to a confession of the same; and upon my indicating the grounds upon which I made the suggestion, permission was granted me. That no time might be lost I entered into discourse with Galileo yesterday afternoon, and, after many arguments and rejoinders had passed between us, by God's grace I attained my object, for I brought him to a full sense of his error, so that he is willing to confess it judicially. He requested, however, a little time in order to consider the form in which he might most fittingly make the confession. I trust his Holiness and your Eminence will be satisfied that in this way the affair will be settled without difficulty; the Court will maintain its reputation; it will be possible to deal leniently with the culprit; and, whatever the decision arrived at, he will recognise the favour shown him, with all the other consequences of satisfaction herein desired. Today I think of examining him in order to obtain the said confession, and, having as I hope received it, it will only remain to me further to question him with regard to his intention, and to impose the prohibitions upon him; and that done he might have the house [Niccolini's] assigned as a prison, as hinted to me by your Eminence.

The second examination did not take place on 28th April as Firenzuola proposed, perhaps on account of Galileo's indisposition. On 30th April the Court again assembled, the usual oath to speak the truth was administered, and Galileo was requested to state what he had to say. He then began the following melancholy confession:

In the course of some days' continuous and attentive reflection on the interrogations put to me on the 12th of the present month, and in particular as to whether sixteen years ago an injunction was intimated to me by order of the Holy Office forbidding me to hold, defend, or teach, in any manner, the opinion that had just been condemned—of the motion of the earth and the stability of the sun—it occurred to me to re-peruse my printed *Dialogues* (which for three years I had not seen), in order carefully to note whether, contrary to my most sincere intention, there had by any inadvertence fallen from my pen anything from which a reader or the authorities might infer not only some taint of disobedience on my part, but also other particulars which might induce the belief that I had contravened the orders of the Holy Church. And being by the kind permission of the authorities at liberty to send about my servant, I succeeded in procuring a copy of my book, and having procured it I applied myself with the utmost diligence to its perusal and to a most minute consideration thereof. And, owing to my not having seen it for so long, it presented itself to me as if it were a new writing and by another author. I freely confess that in several places it seemed to me set forth in such a form that a reader ignorant of my real purpose might have had reason to suppose that the arguments adduced on the false side, and which it was my intention to refute, were so expressed as to be calculated rather to compel conviction by their cogency than to be easy of refutation. Two arguments there are in particular—the one taken from the sun-spots, the other from the ebb and flow of the tide—which in truth come to the ear of the reader with far greater show of force and power than ought to have been imparted to them by one who regarded them as inconclusive and who intended to refute them; as, indeed, I truly and sincerely held and do hold them to be inconclusive and admitting of refutation. And as excuse to myself for having fallen into an error so foreign to my intention, not contenting myself merely with

> saying that when a man recites the arguments of the opposite side with the object of refuting them, he should, especially if writing in the form of dialogue, state them in their strictest form, and should not cloak them to the disadvantage of his opponent. Not contenting myself with saying this, I now see I was misled by that natural complacency which every man feels with regard to his own subtleties and in showing himself more skilful than the generality of men in devising, even in favour of false propositions, ingenious and plausible arguments. However, although with Cicero *avidior sim gloriae quam satis est,* if I had now to set forth the same reasonings, without doubt I should so weaken them that they should not be able to make an apparent show of force of which they are really and essentially devoid. My error then has been—and I confess it—one of vainglorious ambition and of pure ignorance and inadvertence. This is what occurs to me to say with reference to this particular, and what suggested itself to me during the re-perusal of my book.

After making this humiliating declaration, Galileo was allowed to withdraw, and no questions were put to him; but he must have concluded from this silence or other sign that he had not gone far enough in the denial of his inmost convictions; perhaps, this penitent acknowledgment of error and vain-glory was not sufficient, and the Inquisitors would be conciliated by the resolution to publicly correct his error—whatever prompted the impulse, he returned at once to the Court and spoke as follows:

> And in confirmation of my assertion, that I have not held and do not hold as true the opinion which has been condemned, if there shall be granted to me, as I desire, means and time to make a clearer demonstration thereof, I am ready to do so; and there is a favourable opportunity for this, seeing that in the work the interlocutors agree to meet again after a certain time, to discuss

several distinct problems of nature connected with the matter discussed at their meetings. As this affords me an opportunity of adding one or two other ' days,' I promise to take up the arguments already adduced in favour of the said opinion, which is false and has been condemned, and to confute them in such most effectual manner as by the blessing of God will be possible to me. I pray, therefore, this sacred tribunal to aid me in this good resolution, and to enable me to put it into effect.

On the evening of the day on which the second hearing took place, 30th April 1633, Galileo was permitted to return to the Tuscan Embassy, on oath not to leave it, not to hold intercourse with any but the inmates of the house, to present himself before the Holy Office when summoned, and to maintain the strictest silence on the subject of the trial. At this wholly unexpected favour, Niccolini and his household were, as we may imagine, filled with delight.

ON 10TH MAY, Galileo was summoned for the third time before the Inquisition, where the Commissary-General, Firenzuola, informed him that eight days were allowed in which to prepare a defence if he wished to do so; but Galileo at once handed in a paper, from which we may conclude that it was written to order, and under the same extra-judicial pressure as made him write his humiliating confession of 30th April. The greater part of this document is taken up with an explanation why he had not mentioned the prohibition of 1616 when applying for the Imprimatur in 1630—which explanation amounts to a formal admission that on 26th February 1616 he was not only commanded not to hold or defend the Copernican doctrine, but not to teach it in any way whatsoever. Coming then to the last paragraph, he says:

Lastly, it remains for me to pray you to take into consideration

> my pitiable state of bodily indisposition, to which at the age of seventy years, I have been reduced by ten months of constant mental anxiety, and the fatigue of a long and toilsome journey at the most inclement season, together with the loss of the greater part of the years of which, from my previous condition of health, I had the prospect. I am encouraged to ask this indulgence by the clemency and goodness of the most eminent lords, my judges, and hope that they will be pleased to remit what may appear good to their entire justice, and to consider my sufferings as adequate punishment.

This touching appeal to the mercy of his judges cannot be read without feelings of the profoundest pity for the crushed old man, who in the evening of a glorious life was thus compelled to deny his inmost convictions, and to sue cravenly for that pity which was not to be shown him.

After his paper had been received, and the same obligations imposed on him on oath as after the second hearing, he was allowed to return to the Embassy. The nearer the time approached when his illusions were to be dispelled, the more sanguine was the intelligence he sent to his friends. A favour granted just at the last, on the urgent solicitation of Niccolini, and unheard of in the annals of the Inquisition, might have encouraged these confident hopes. He was permitted to take the air in the gardens of the Villa Medici on the Pincio, to which, however, he was always conveyed in a closed carriage, as he must not be seen in the streets! Niccolini did not share in these hopes of his guest. After an audience of the Pope and Cardinal Barberini, he wrote to Cioli (22nd May):

> I very much fear that the book will be prohibited, unless it is averted by Galileo's being charged (as I suggested) to write an apology. Some salutary penance will also be imposed, as they maintain that he has transgressed the command given to him by

> Cardinal Bellarmine in 1616. I have not yet told Galileo all this, because I want to prepare him for it by degrees, in order not to distress him.

A lull now took place in the proceedings—the preparation for the great catastrophe that was to crush Galileo and fill the educated world with horror. Sultry silence reigned for four weeks, and no one, not even Niccolini, could learn anything. Indeed, the thunderbolt had fallen before the fact was known outside the Holy Office. Galileo's fate had been sealed at a private meeting of the Congregation, 16th June 1633, at which the Pope presided. It was decreed to try Galileo as to his intention, under threat of torture, and if this failed he was then to be called upon to recant before a plenary assembly of the Holy Office; to be condemned to imprisonment at their pleasure; and to be ordered in future not to discuss in writing or speaking the opinion that the earth moves and that the sun is stationary, nor even the contrary opinion, under pain of further punishment as a relapsed heretic. Further, the work "Dialogo di Galileo Galilei, Linceo," was to be prohibited. And, in order to make this known everywhere, copies of the sentence were to be sent to all Papal envoys and to all Inquisitors into heretical crimes, and especially to the Inquisitor in Florence, who were to read it publicly to all professors of mathematics summoned for the purpose.

TWO DAYS after these proceedings had been determined on, the Pope received Niccolini, who once more came to beg for a speedy termination of the trial, the long suspense of which was torturing to him only in a less degree than it was to Galileo himself. Urban coolly replied that it had already been terminated, and that within the next few days Galileo would be summoned to hear his sentence. The Ambassador, aghast at this unexpected answer, implored the Pope to mollify any severity that the judges might, perhaps, have thought necessary, and added that the great complaisance hitherto shown to his

Sovereign's wishes in the matter was fully appreciated, and that the Grand Duke was only waiting the end of the business to express his gratitude in person. The Pope replied that his Highness need not take this trouble, that he had readily granted every amelioration possible; but as to Galileo's opinions they could do no less than prohibit them as erroneous and contrary to Holy Scripture, and as to his person, he would, according to usage, be imprisoned for a time, because he had transgressed the mandate issued to him in 1616. "However," added Urban, "after the publication of the sentence we will see you again and we will consult together so that he may suffer as little distress as possible."

The same day, 18th June, Niccolini reported this audience to Cioli, and remarked at the end that he had simply informed Galileo of the approaching end of the trial and of the prohibition of his book, but had said nothing about the personal punishment, in order not to trouble him too much at once. The Pope had also enjoined this course, "because, perhaps, in the course of the proceedings things might take a better turn." The drama now rapidly proceeded to a climax. On the evening of 20th June 1633, Galileo was warned to appear before the Inquisition on the following morning, when, as we know from the programme of 16th June, he was to be questioned under threats of torture about his "intention," that is, as to his real convictions.

On the morning of the 21st, Galileo appeared before his judges. After he had taken the usual oath and had answered in the negative the query whether he had any statement to make, the examiner asked three separate questions, slightly varied but all to the effect whether he has held and holds the opinion that the sun is the centre of the world, and that the earth is not the centre, but moves, and with a diurnal motion. To these Galileo suitably replied in the negative. "I do not hold," he says, in reply to the last query, "and have not held this opinion of Copernicus since the command was given me that I must abandon it. For the rest I am here in your hands, do with me as you please." Being once more bidden to speak the truth, otherwise recourse would be had to torture, the terrified old man answered with the resignation of despair, "I am here to obey. I have not held this opinion since the decision was pronounced, as I have stated."

In the protocol of the trial the concluding sentence follows immediately after Galileo's answer: "And as nothing further could then be done in execution of the decree [of 16th June] his signature was obtained to his depositions, and he was sent back to his place," that is, to some place in the buildings of the Holy Office, where he was detained till 24th June.

~~ST. PETER DENYING CHRIST~~

GALILEO DENYING SCIENCE

And Galileo sat and defended himself. They said therefore unto him, "Art not thou also one of Copernicus' disciples?" He denied it, and said, "I am not."

13

SENTENCE OF THE INQUISITION—EXILED TO SIENA

1633

On Wednesday, 22nd June 1633, in the forenoon, Galileo was conducted to the large hall used for melancholy proceedings of this kind in the Dominican Convent of Santa Maria sopra Minerva, where in the presence of his judges and a large assemblage of cardinals and prelates of the Church his sentence was read to him as follows:

We, the undersigned,

GASPARO of Santa Croce in Gerusalemme, Borgia,
FRA FELICE CENTINO of S. Anastasia, called Ascoli,
GUIDO of Santa Maria del Popolo, Bentivoglio,
FRA DESIDERIO SCAGLIA of S. Carlo, called Cremona,
FRA ANTONIO BARBERINI, called S. Onofrio,
LAUDIVIO ZACCHIA of S. Pietro in Vincoli, called San-Sisto,
BERLINGERO of San Agostino, Gessi,
FABRIZIO of S. Lorenzo in Pane e Perna, Verospi,
FRANCESCO of S. Lorenzo in Damaso, Barberini,
MARTINO of Santa Maria Nuova, Ginetti,
by the grace of God, Cardinals of the Holy Roman Church,

Inquisitors General throughout the whole Christian Republic, Special Deputies of the Holy Apostolical Chair against heretical depravity:

Whereas you, Galileo, son of the late Vincenzio Galilei, of Florence, aged 70 years, were denounced, in 1615, to this Holy Office, for holding as true a false doctrine taught by many, namely, that the sun is immovable in the centre of the world, and that the earth moves, and also with a diurnal motion; also, for having pupils whom you instructed in the same opinions; also, for maintaining a correspondence on the same with some German mathematicians; also for publishing certain letters on the sun-spots, in which you developed the same doctrine as true; also for answering the objections which were continually produced from the Holy Scriptures, by glossing the said Scriptures according to your own meaning; and whereas thereupon was produced the copy of a writing, in form of a letter, professedly written by you to a person formerly your pupil, in which, following the hypothesis of Copernicus, you include several propositions contrary to the true sense and authority of the Holy Scriptures; therefore (this Holy Tribunal being desirous of providing against the disorder and mischief which were thence proceeding and increasing to the detriment of the Holy Faith) by the desire of his Holiness and of the Most Eminent Lords, Cardinals of this supreme and universal Inquisition, the two propositions of the stability of the sun, and the motion of the earth, were qualified by the Theological Qualifiers as follows:

1. The proposition that the sun is in the centre of the world and immovable from its place is absurd, philosophically false, and formally heretical; because it is expressly contrary to the Holy Scriptures.

2. The proposition that the earth is not the centre of the world, nor immovable, but that it moves, and also with a diurnal motion, is also absurd, philosophically false, and, theologically considered, at least erroneous in faith.

But whereas, being pleased at that time to deal mildly with

you, it was decreed in the Holy Congregation, held before his Holiness on the twenty-fifth day of February 1616, that his Eminence the Lord Cardinal Bellarmine should enjoin you to give up altogether the said false doctrine; and if you should refuse, that you should be ordered by the Commissary of the Holy Office to relinquish it, not to teach it to others, nor to defend it; and in default of acquiescence, that you should be imprisoned; and whereas in execution of this decree, on the following day, at the Palace, in presence of his Eminence the said Lord Cardinal Bellarmine, after you had been mildly admonished by the said Lord Cardinal, you were commanded by the Commissary of the Holy Office, before a notary and witnesses, to relinquish altogether the said false opinion, and, in future, neither to defend nor teach it in any manner, neither verbally nor in writing, and upon your promising obedience you were dismissed.

And, in order that so pernicious a doctrine might be altogether rooted out, nor insinuate itself further to the heavy detriment of the Catholic truth, a decree emanated from the Holy Congregation of the Index prohibiting the books which treat of this doctrine, declaring it false, and altogether contrary to the Holy and Divine Scripture.

And whereas a book has since appeared published at Florence last year, the title of which showed that you were the author, which title is *The Dialogue of Galileo Galilei, on the two principal Systems of the World—the Ptolemaic, and Copernican;* and whereas the Holy Congregation has heard that, in consequence of printing the said book, the false opinion of the earth's motion and stability of the sun is daily gaining ground, the said book has been taken into careful consideration, and in it has been detected a glaring violation of the said order, which had been intimated to you; inasmuch as in this book you have defended the said opinion, already, and in your presence, condemned; although, in the same book, you labour with many circumlocutions to induce the belief that it is left by you undecided and merely probable;—

which is equally a very grave error, since an opinion can in no way be probable which has been already declared and finally determined contrary to the Divine Scripture. Therefore, by Our order, you have been cited to this Holy Office, where, on your examination upon oath, you have acknowledged the said book was written and printed by you. You also confessed that you began to write the said book ten or twelve years ago, after the order aforesaid had been given. Also, that you had demanded licence to publish it, without signifying to those who granted you this permission that you had been commanded not to hold, defend, or teach, the said doctrine in any manner. You also confessed that the style of the said book was, in many places, so composed that the reader might think the arguments adduced on the false side to be so worded as more effectually to compel conviction than to be easily refutable, alleging, in excuse, that you had thus run into an error, foreign (as you say) to your intention, from writing in the form of a dialogue, and in consequence of the natural complacency which every one feels with regard to his own subtleties, and in showing himself more skilful than the generality of mankind, in contriving, even in favour of false propositions, ingenious and plausible arguments.

And, upon a convenient time being given you for making your defence, you produced a certificate in the handwriting of his Eminence the Lord Cardinal Bellarmine, procured, as you said, by yourself, that you might defend yourself against the calumnies of your enemies, who reported that you had abjured your opinions, and had been punished by the Holy Office; in which certificate it is declared that you had not abjured nor had been punished, but merely that the declaration made by his Holiness, and promulgated by the Holy Congregation of the Index, had been announced to you, which declares that the opinion of the motion of the earth and stability of the sun is contrary to the Holy Scriptures, and, therefore, cannot be held or defended. Wherefore, since no mention is there made of two articles of the order, to wit, the order 'not to teach' and 'in any manner,' you argued that we ought to believe that, in the lapse of fourteen or sixteen years, they had escaped your memory, and that this was also the reason why you were silent as to the order when you sought permission to publish your book, and that this is said by you, not to excuse your error, but that it may be attributed to vain-glorious ambition rather than to malice. But this very certificate, produced on your behalf, has greatly aggravated your offence, since it is therein declared that the said opinion is contrary to the Holy Scriptures, and yet you have dared to treat of it, and to argue that it is probable. Nor is there any extenuation in the licence artfully and cunningly extorted by you, since you did not intimate the command imposed upon you. But whereas it appeared to Us that you had not disclosed the whole truth with regard to your intention, We thought it necessary to proceed to the rigorous examination of you, in which (without any prejudice to what you had confessed, and which is above detailed against you, with regard to your said intention) you answered like a good Catholic.

Therefore, having seen and maturely considered the merits of your cause, with your said confessions and excuses, and everything else which ought to be seen and considered, we have come to the underwritten final sentence against you:

Invoking, therefore, the most holy name of our Lord Jesus Christ, and of His Most Glorious Virgin Mother, Mary, We pronounce this Our final sentence, which, sitting in council and judgment with the Reverend Masters of Sacred Theology and Doctors of both Laws, Our Assessors, We put forth in this writing in regard to the matters and controversies between the Magnificent Carlo Sincero, Doctor of both Laws, Fiscal Proctor of the Holy Office, of the one part, and you, Galileo Galilei, defendant, tried and confessed as above, of the other part. We pronounce, judge, and declare, that you, the said Galileo, by reason of these things which have been detailed in the course of this writing, and which, as above, you have confessed, have rendered yourself vehemently suspected by this Holy Office of heresy, that is of having believed and held the doctrine (which is false and contrary to the Holy and Divine Scriptures), that the sun is the centre of the world, and that it does not move from east to west, and that the earth does move, and is not the centre of the world; also, that an opinion can be held and supported as probable, after it has been declared and finally decreed contrary to the Holy Scripture, and, consequently, that you have incurred all the censures and penalties enjoined and promulgated in the sacred canons and other general and particular constitutions against delinquents of this description. From which it is Our pleasure that you be absolved, provided that with a sincere heart and unfeigned faith, in Our presence, you abjure, curse, and detest, the said errors and heresies, and every other error and heresy, contrary to the Catholic and Apostolic Church of Rome, in the form now shown to you.

But that your grievous and pernicious error and transgression may not go altogether unpunished, and that you may be made more cautious in future, and may be a warning to others to abstain from delinquencies of this sort, we decree that the book *Dialogues of Galileo Galilei* be prohibited by a public edict, and We condemn you to the formal prison of this Holy Office for a period determinable at Our pleasure; and by way of salutary penance, We order you during the next three years to recite, once a week, the seven penitential psalms, reserving to Ourselves the power of moderating, commuting, or taking off, the whole or part of the said punishment or penance.

And so We say, pronounce, and by Our sentence declare, decree, and reserve, in this and in every other better form and manner, which lawfully We may and can use. So We, the subscribing Cardinals, pronounce.

FELIX, Cardinal di Ascoli.

GUIDO, Cardinal Bentivoglio.

DESIDERIO, Cardinal di Cremona.

ANTONIO, Cardinal S. Onofrio.

BERLINGERO, Cardinal Gessi.

FABRIZIO, Cardinal Verospi.
MARTINO, Cardinal Ginetti.

IN CONFORMITY with the foregoing sentence, Galileo was made to kneel before the Inquisition, and make the following abjuration:

I, Galileo Galilei, son of the late Vincenzio Galilei of Florence, aged 70 years, being brought personally to judgment, and kneeling before you, Most Eminent and Most Reverend Lords Cardinals, General Inquisitors of the Universal Christian Republic against heretical depravity, having before my eyes the Holy Gospels which I touch with my own hands, swear that I have always believed, and, with the help of God, will in future believe, every article which the Holy Catholic and Apostolic Church of Rome holds, teaches, and preaches. But because I have been enjoined, by this Holy Office, altogether to abandon the false opinion which maintains that the sun is the centre and immovable, and forbidden to hold, defend, or teach, the said false doctrine in any manner; and because, after it had been signified to me that the said doctrine is repugnant to the Holy Scripture, I have written and printed a book, in which I treat of the same condemned doctrine, and adduce reasons with great force in support of the same, without giving any solution, and therefore have been judged grievously suspected of heresy; that is to say, that I held and believed that the sun is the centre of the world and immovable, and that the earth is not the centre and movable, I am willing to remove from the minds of your Eminences, and of every Catholic Christian, this vehement suspicion rightly entertained towards me, therefore, with a sincere heart and unfeigned faith, I abjure, curse, and detest the said errors and heresies, and generally every other error and sect

> contrary to the said Holy Church; and I swear that I will never more in future say, or assert anything, verbally or in writing, which may give rise to a similar suspicion of me; but that if I shall know any heretic, or any one suspected of heresy, I will denounce him to this Holy Office, or to the Inquisitor and Ordinary of the place in which I may be. I swear, moreover, and promise that I will fulfil and observe fully all the penances which have been or shall be laid on me by this Holy Office. But if it shall happen that I violate any of my said promises, oaths, and protestations (which God avert), I subject myself to all the pains and punishments which have been decreed and promulgated by the sacred canons and other general and particular constitutions against delinquents of this description. So, may God help me, and His Holy Gospels, which I touch with my own hands, I, the above named Galileo Galilei, have abjured, sworn, promised, and bound myself as above; and, in witness thereof, with my own hand have subscribed this present writing of my abjuration, which I have recited word for word.
>
> At Rome, in the Convent of Minerva, 22nd June 1633, I, GALILEO GALILEI, have abjured as above with my own hand.

After this sorrowful drama had been concluded, Galileo was led back to the buildings of the Holy Office. And now that he and the Copernican system had been condemned with all the terrifying forms of the Inquisition, Urban's wounded vanity was soothed, and he was pleased to give the word for a little mercy. Galileo was not, as the sentence prescribed, to be detained in the prison of the Holy Office, but was banished to the villa of the Grand Duke of Tuscany at Trinita dei Monti, which, by Papal orders dated 23rd June, he was to consider as a prison. Thither, where, many years before, he had shown the moons of Jupiter and other "Celestial Novelties" to wondering cardinals, he was now conducted by the ever-faithful Niccolini on the evening of 24th June.

From Niccolini's letter to Cioli of 26th June we learn that while Galileo took the prohibition of his book (for which his friend had prepared him) with tolerable composure, the wholly unexpected proceedings against himself personally affected him terribly. He sank into a deep depression from which the Ambassador did his best to rouse him, but with little success for a time.

IN ACCORDANCE with the decree of 16th June 1633, copies of the sentence and abjuration were despatched to all Papal Nuncios, all Inquisitors, and many Universities, Italian and Foreign. In Padua and Florence especially, the means of publication were calculated with a refinement of cruelty. In Padua the local Inquisitor read both documents to the professors of philosophy and mathematics, and to the students convened for the purpose in the University library. A search was made for copies of the condemned book, and if none were found the Inquisitor could at least boast of one voluntary surrender. The peripatetic Fortunio Liceti gave up his copy—a presentation one by the author! In Florence the Inquisitor read the sentence and abjuration publicly in the church of Santa Croce, notices to attend having been previously served on all professors, and on all others who were known to be friends and adherents of Galileo. Thus Aggiunti, Guiducci, Arrighetti, and many others who loved the great master were made to participate in his humiliation.

But the cup of Papal wrath was not emptied on Galileo's head alone. All who had befriended him, or had any part in the licensing of his *Dialogues*, were punished in some way. Ciampoli, in December 1632, was deprived of his office of Secretary of the Papal Briefs and was (practically) exiled as Governor of Montalto, where he remained as long as he lived—Galileo's faithful disciple to the last. Early in April 1633 Galileo was deprived of the valuable advocacy of the devoted Castelli, who was sent away on some pretext, and was not recalled until the middle of 1635, when "at last he had again been permitted to kiss his Holiness's toe." The Inquisitor in Florence was severely reprimanded, and Riccardi, the Censor, was dismissed in disgrace and deprived of all offices during Urban's lifetime.

The inconsistency of these proceedings will be noted in the latter cases. These people are punished for granting that very licence which Galileo was charged with, and condemned for, having surreptitiously obtained from them by concealing circumstances with which they were not bound to be acquainted. Riccardi, in exculpation of his conduct, produced a letter from Ciampoli, in which it was said that his Holiness (in whose presence the letter professed to be written) ordered the licence to be given; but the Pope only replied that this was a Ciampolism; that his Secretary and Galileo had circumvented him; that he had already dismissed Ciampoli, and that Riccardi must prepare to follow him.

~~THE DARKNESS AT THE CRUCIFIXION~~

THE DEATH OF KNOWLEDGE

And it was about the year 1633, and there was a darkness over all the earth until the year 1835.

GALILEO, after his first great anguish had somewhat subsided, felt that he must quit Rome and its hateful memories, and so addressed a pitiable letter to the Pope.

"Most Holy Father," he says, "Galileo Galilei humbly begs your Holiness to exchange the place assigned to him for his prison near Rome for some other in Florence, which may appear suitable to your Holiness, in consideration of his poor health, and also because he is expecting a sister [in law] with eight children from Germany, to whom no one can afford help and protection so well as himself. He will receive any disposition of your Holiness as a great favour."

But at the Vatican it was thought that to allow Galileo to return at once to Florence would be a superfluity of indulgence. "We must proceed gently," said his Holiness, "and only rehabilitate him by degrees." Still Urban was moved to some compassion, and on 30th June allowed the poor old man to retire to Siena, to the house of his former pupil Archbishop Ascanio Piccolomini, where he was to remain under the orders of the Archbishop, and on no account to leave the town without permission from Rome. Galileo was informed of this decision on 2nd July, and early on the 6th he shook the dust of Rome from off his feet. Niccolini, reporting his departure to Cioli (on 10th July), says: "Signor Galileo set out early on Wednesday in good health for Siena, and writes to me from Viterbo that he had performed four miles on foot, the weather being very cool."

Galileo reached Siena safely on 9th July, and was warmly received by Piccolomini; but neither his devoted kindness, not the stimulating converse with an old friend, the learned Alessandro Marsili, then residing at Siena, could make him forget that he was a prisoner of the Inquisition.

On 23rd July 1633, he wrote to Cioli:

Today I address myself to you, oppressed by the ennui of a

> captivity of more than six months—a captivity made more painful by the chagrin and the anxiety of the preceding year, and by all the dangers and all the bodily sufferings which have followed in their train. My misfortunes are commiserated by all the world, except by those who have judged me deserving of this punishment—but of this another time.
>
> The duration of my captivity is entirely at the pleasure of his Holiness. On the intervention of the Ambassador Niccolini, the Pope assigned me, instead of the prison of the Holy Office, the Villa de Medici, Trinita dei Monti, where I remained some days. Again, on the Ambassador's intercession, I was sent to the Archiepiscopal Palace here, where for fifteen days I have experienced the greatest kindness from the excellent Archbishop. But, apart from the wish that I have to return to my home and to be at liberty, this liberty is really essential to me. Therefore, I beg you to move his Highness to solicit the favour of my liberty from his Holiness, or through Cardinal Barberini. You might point out that the house of the Grand Duke has been for a long time deprived of my services, and insist on attaching to this circumstance more importance than it merits.

After all his bitter disillusions we see that his hopefulness had not yet abandoned him. The Grand Duke very kindly consented to exert once more what influence he possessed with the Pope, and instructed his ambassador accordingly; but Niccolini represented that the moment was not opportune, and recommended that action should be deferred for a few months. Meanwhile the good Archbishop did all that love for his venerated guest could suggest to make his house as little like a prison as possible, rather "an earthly paradise," as Maria Celeste wished her father to consider it; but for a long time these efforts had little effect. The cruel edict condemning him to perpetual silence on a subject which was one of the mainsprings of his life was a serpent's sting which could not be readily forgotten; his soul was lacerated, and he fell into frequent fits of utter despondency, in which he accused his friends of having forgotten him. In one of these moments of bitterness he wrote to his daughter—"my name is erased from the book of the living." "Nay," came Maria Celeste's soothing reply, "say not that your name is struck out *de libro viventium,* for it is not so, neither in the greater part of the world, nor in your own country. Indeed, it seems to me that if for a brief moment your name and fame were clouded they are now restored to greater brightness; at saying which I am much astonished with myself, for I know that generally *nemo propheta acceptus in patria sua.* I am afraid that if I go on quoting Latin I shall fall into some barbarism, so I shall stop. But, indeed, you are loved and esteemed here more than ever."

As the weary months rolled on, Galileo became a little resigned to his situation at Siena, and even began to occupy himself with another of his great works (indeed the greatest), namely *Dialoghi delle Nuove Scienze* (or, *Dialogues on The New Sciences*), the writing of which he spoke of as far back as 1610, in his letter to Vinta. His interest in other Scientific matters was as keen as ever. Thus, writing (27th September) to Andrea Arrighetti, a young Florentine disciple who had sent him some mathematical problems, he says:

> The pleasure with which I read and re-read your demonstrations was greater than my astonishment, since the pleasure was proportionate to the sagacity of which you give proof in your argumentation, while the astonishment was little, because I remembered I had under my eyes a work of Signor Andrea Arrighetti.
>
> The last theorem held me for a moment in meditation and in doubt, as much owing to the unusual formula as to fatigue of memory, which lets escape impressions as soon as formed. Let this be a lesson to you, and encourage you to exercise the mind while you are young.
>
> As regards myself, I can say that my relations with my kind and honoured host bring me much consolation, and in the midst of so many sad subjects for meditation they give a new direction

> to my thoughts. But more than any other consolation, the knowledge that you and my other friends retain for me your old affection makes my grief less heavy.

Galileo's detention at Siena would, perhaps, have been borne more easily did he not know that his loved and loving daughter, in spite of her resignation, was consuming her poor heart with longing to see him once more. "When you were in Rome I said to myself, if he were but at Siena! Now you are at Siena, I say, would he were at Arcetri! But God's will be done!" Her life was one continual prayer for him. Yet, while ever thinking of his spiritual welfare, she did not neglect his worldly affairs. In her letters of this period she tells him of the fruit and the wine which have been sold; of the incoming and outgoing of his money; that the vines had been injured by hail; that thieves had been over the garden wall; that his mule was behaving badly and would carry no one now her master was away; that a storm had damaged the roof, and thrown down and smashed a large vase; that the plums were few; and that the wind had carried away the pears; and so on. With the money from the sale of some lemons she had had three masses said for her father's special benefit. Finally: "There are two pigeons," she says, "in the dovecote waiting for you to come and eat them; there are beans in the garden waiting for you to gather them; and your tower is lamenting your long absence."

As soon as the quarantine regulations were relaxed, Maria Celeste sent the boy Geppo on the mule to Siena to bring back news of her father, how he was looking, etc. The poor old man seems to have asked her to remember him in her prayers, for on Geppo's return she wrote (3rd October):

> It seems to me a thousand years till I see you back again safe and well. I would not have you doubt that all this time I have never ceased from commending you to God with my whole heart, for, indeed, I feel too anxious for your spiritual and bodily health ever

> to have neglected praying for you. To give you a proof I will tell you that as a great favour I had a copy of your sentence shown to me, and though on the one hand it grieved me to read it, yet on the other hand I was glad, because I found out a way of being of some slight use to you, namely, by taking on myself that part of the sentence which orders you to recite the seven Penitential Psalms once a week. I began to do this a while ago, and it gives me much pleasure—first, because I am persuaded that prayer in obedience to the Holy Church must be efficacious; secondly, in order to save you the trouble of remembering it. If I had been able to do more, most willingly would I have entered a straiter prison than the one I live in now, if by so doing I could set you at liberty.

At length, the weariness and sickness of heart caused by hope deferred began to tell on this sweet nun. Worn by continual ill-health, by anxiety for her father, by nightly watchings in the convent infirmary, and daily occupations in the stillroom and pharmacy, she would appear to have felt a presentiment of her approaching dissolution. She strove gently to prepare her father, telling him it was for him to live long to the service and the glory of the God who had endowed him with such a wondrous intellect, and to the comfort of many who would feel his loss. As for herself, she could neither do much for the glory of God, nor be of much use to any one, and her living or dying would make little difference.

IN NOVEMBER 1633, thinking the time favourable, Niccolini began to agitate for Galileo's pardon, but the Pope was not disposed to go so far, and pretended there would be a difficulty in getting the Congregation of the Holy Office to consent to such a course—a patent evasion, as the decision rested solely and entirely with himself. Niccolini, however, persisted in his efforts, and went to Cardinal Barberini and other members of the Congregation to enlist their good offices. At length, on 1st December, the question of Galileo's pardon, or rather release from personal restraints, came before the Congregation—the Pope presiding—and, though recommended by Cardinal Barberini, it was refused; but Galileo was permitted to retire to his villa at Arcetri, where he was to remain till further orders, and where he might receive his friends and relations, but not too many at a time, if he wished to avoid suspicion.

While Niccolini's letter of 3rd December, containing this piece of good news, was on its way to Siena, Galileo was writing thus to Bocchineri in Florence (9th December):

> For the last four days I have suffered from violent pains in the limbs which are more persistent than ordinary. I fear greatly that this climate (much more rigorous in winter than that of Florence) is the principal cause of these ailments; and I foresee that I shall be very seriously crippled if obliged to remain here much longer. I await a decision from Rome, but I have little hope of its being favourable.

For once Galileo despaired, and at the wrong moment! The same day Niccolini's letter of 3rd December arrived, and a few days afterwards he set out for Arcetri.

Then Galileo's students went to Arcetri, to the house where Galileo had been confined. When they saw him, they praised his intellect; but some doubted his theories. Then Galileo came to them and said:

"All knowledge regarding the heavens and the earth has been given to me. Therefore go and make students of all nations, recruiting them in the name of Science and of Technology and of the Spirit of Progress, and teaching them everything I have instructed you. And surely my insights will be with you always, to the very end of the universe."

~~THE RESURRECTION~~

THE EMPTY LIBRARY

And the messenger answered and said unto the students, "Fear not ye: for I know that ye seek Knowledge, which was incinerated. Knowledge is not here: for it will rise again, as Galileo said. Come, see the place where Knowledge once accumulated."

THE GOSPEL OF ISAAC NEWTON [SAMPLE]

1

EARLY LIFE AND EDUCATION

The name of this celebrated philosopher has by general consent been placed at the head of those great men who have been acknowledged as the ornaments of their species. The character of Newton cannot be delineated and discussed like that of ordinary men; its unity is so beautiful, that the biographer must dwell upon it with delight, and the inquiry by what means he attained an undisputed superiority over his fellow-creatures must be both interesting and useful. It has been asserted that all men are born equal in talents, and that the difference which exists amongst them is the effect of education; but this is disproved by the observation of every parent and teacher—a decided inequality in capacity for receiving instruction being distinctly exhibited by children even in infancy. Newton was endowed with talents of the highest order, but those who are less eminently gifted may study his life with advantage, and derive instruction from every part of his career. With a power of intellect almost divine he demonstrated the motion of the planets, the orbits of the comets, and the causes of the tides of the ocean; he investigated with complete success the proportions of light and colours, which no man before had even suspected; he was the diligent, sagacious, faithful interpreter of Nature, antiquity, and Scripture. He was deeply learned in history and chronology, and he applied the unrivalled powers of his mighty intellect to the complete examination of a subject compared with which all others sink into insignificance; the result was a clear conviction of the truth of revealed religion, which is demonstrated in all his works, and which was still more effectually shown in his life and conduct. Those who consider the character of an individual so highly renowned will duly appreciate the value of his testimony. His biography, therefore, cannot fail to excite a general interest.

ST. ISAAC NEWTON was born at Woolsthorpe, a hamlet in the parish of Colsterworth, in Lincolnshire, about six miles south of Grantham, on the 25th day of December (old style) 1642, and was baptised at Colsterworth on the 1st day of January 1642–3. His father, Mr. Isaac Newton, died at the early age of thirty-six, only a few months after his marriage to Harriet Ayscough, daughter of James Ayscough of Market Overton in Rutlandshire. The consequence was, that the mother of the interesting subject of these pages was left in a state of pregnancy, and appears, from the grief occasioned by the death of her husband, to have given birth to her first and posthumous child. The helpless infant thus ushered into the world was of such an extremely diminutive, and seemed of so delicate a frame, that two women who were hurried off to Lady Pakenham's at North Witham, to procure some medicine for the purpose of strengthening the puling infant, declared that they did not expect to find him in life on their return. St. Isaac in later life repeatedly asserted that he had often heard his mother say, that "when he was born he was so little that they might have put him into a quart mug, and did not expect that he could live." But an all-wise Providence had otherwise decreed, and that feeble body which appeared scarcely qualified to contain its immortal mind, was destined to shoot forth and enjoy a vigorous maturity, and to survive the threescore years and ten allotted as the term of human existence. The property of Woolsthorpe, in the manor-house of which this remarkable birth took place, has been in the possession of the family of Newton for more than a hundred years. This family came originally from Newton in Lancashire, but, before purchasing the estate of Woolsthorpe, had settled at Westby in Lincolnshire. The manor-house is situated in a beautiful little valley, celebrated for its copious wells of excellent spring water, on the west side of the river Witham, which takes its rise in the vicinity, and commands a remarkably pleasing prospect to the east towards Colsterworth. The paternal property of Woolsthorpe was worth no more than £30 a-year; but Mrs. Newton was proprietor of another small estate at Sewstern, in Leicestershire and only three miles distant from Woolsthorpe, which raised the yearly value of their property to nearly £80; and being a woman of an active disposition, it is probable that the cultivation of the small farm on which she resided somewhat increased the limited income upon which she had to support herself and her favourite child, as well as to provide for his education.

Mrs. Newton continued to nurture and watch over her tender charge for three years, with the most exemplary parental solicitude; but at the end of that period, in consequence of her marriage to the Reverend Barnabas Smith, rector of North Witham, she removed to the parsonage, and Isaac was left under the care of his maternal grandmother. At an early age he was sent to two day schools at Skillington and Stoke, where he acquired the little learning usually afforded at these seminaries; but upon reaching his twelfth year he went to the public school at Grantham, conducted by Mr. Stokes, and was at the house of Mr. Clark, an apothecary in that town. At this time he was a thoughtless boy, very inattentive to his lessons, and much attached to boyish amusements, and consequently very low in the school, and presented little hopes of his future success. The boy, however, who happened to be above him in the class, one day gave him a severe kick upon his stomach, from which he suffered great pain. Isaac's mode of revenge was peculiar; from his little size, and delicate constitution, he was unable to punish his antagonist in the usual method, by giving him a good beating; but he laboured incessantly at his studies, and gave up his amusements, till he got above him in the school; and from that period he rose by degrees, till he was the head boy. From the industrious habits of application which this trifling incident had led him to form, the peculiar character of mind of the future philosopher was speedily displayed.

During the hours of play, when his companions were occupied with these amusements, his attention was taken up with mechanical contrivances, either in imitation of something which he had previously observed, or in the execution of some original conception of his own. To enable him to do so, he embraced every opportunity of providing himself with little saws, hatchets, hammers, and all sorts of tools, which, by constant practice, he soon acquired the art of using with a workmanlike dexterity. The principal pieces of mechanism which he thus constructed at this early age were a windmill, a water-clock, and a carriage put in motion by the person who sat in it. When a windmill was in the course of erection in the neighbourhood of Grantham, on the road to Gunnerly, young Newton frequently attended and anxiously watched the operations of the workmen, and attained such a complete knowledge of the machinery, that in a short time he completed a working model of it, to the great surprise and admiration of every person who beheld it. The young mechanist frequently placed this model on the top of Mr. Clarke's house, in which he lodged, where it was put in motion by the action of the wind upon its sails; but not content with this exact representation of the original machine, he conceived the idea of driving it by animal power; and for this purpose he entrapped a mouse, which he called the miller; and this little animal, by acting on a sort of tread-wheel, gave motion to the machine. According to some accounts, the mouse was made to advance by having a string tied to its tail, while others allege that the power of the diminutive agent was instigated by its fruitless attempts to reach a small piece of toasted cheese placed above the wheel.

His water-clock was formed out of a box, which, with great earnestness, he had solicited from the brother of his landlady, without giving any notice of his intentions. It was about four feet high, and of proportionate breadth, not unlike a common house-clock. The index of the dial-plate was turned by a piece of wood, which either rose or fell by the action of dropping water. As it stood in Isaac's own bedroom, he supplied it every morning with the requisite quantity of water; and this same machine was used as a time-keeper by Mr. Clarke's family, and remained in the house many years after its inventor had left Grantham. Dr. Stukely, in a letter to Dr. Mead, says, "I remember once, when I was deputy to Dr. Halley, Secretary to the Royal Society, St. Isaac talked of these kind of instruments. That he observed the chief inconvenience in them was, that the hole through which the water was transmitted was necessarily very small, and subject to be furred up by the impurities in the water, as those made with sand will wear bigger, which at length causes an inequality in time." This satisfactorily proves that Newton's mind was occupied by the consideration of his youthful inventions, when more important avocations attracted his attention in after years.

His mechanical carriage was a vehicle with four wheels, which was put in motion with a handle wrought by the person who sat in it, but it seems only to have been calculated for a smooth surface, and not fitted to overcome the inequalities of a road. It was, however, a most useful invention for invalids, which enabled them to move from one place to another in a room, when unable to walk.

ISAAC NEWTON WAS at this period "a sober, silent, thinking lad" who seldom ever joined in the ordinary games and more boisterous amusements of his companions, yet he took great delight in providing them with pastimes of a Scientific character. He introduced among his schoolfellows the flying of paper kites, and he is said to have been at great pains in determining their best forms and proportions, and in ascertaining the position and number of the points by which the string ought to be attached. He also made paper lanterns, by the light of which he went to school in the winter mornings; and in dark nights he frequently attached these lanterns to the tails of his kites, which inspired the country people with the belief that they were comets.

In the house of Mr. Clark at Grantham, where he lodged, there were also some female boarders in whose company he appears to have taken much pleasure. One of these young ladies, a Miss Storey, sister to Dr. Storey, a physician at Buckminster, near Colsterworth, was Newton's junior by two or three years, and to great personal attractions she seems to have added more than the customary share of female talent. The society of Miss Storey and her companions was always preferred to that of his own schoolfellows, and it was one of his most agreeable occupations to exercise his mechanical abilities in constructing for them little tables, chairs, cupboards, and other utensils for holding their dolls and their trinkets. He had lived nearly six years in the same house with Miss Storey, and there is every reason to believe, that during this time their youthful friendship gradually rose to a stronger passion; and his attachment is said to have continued even after he was sent to college, but as he could not marry without forfeiting his chance of a fellowship, and as he had no means of supporting a wife and family, he subdued his predilection in silence, and found consolation in the severest labour of study. This young lady was afterwards twice married, and under the name of Mrs. Vincent, Dr. Stukely visited her at Grantham in the year 1727, when she had attained the mature age of eighty-two, and from her he obtained many interesting particulars respecting the early history of St. Isaac. Newton's esteem for this his first love continued unabated during his life. He never visited Lincolnshire without calling upon her; and, when she sunk into poverty, liberally supplied her wants, by relieving her from little pecuniary difficulties which appear to have frequently perplexed her. But Newton's genius was not engrossed by mechanical pursuits. Among his early passions we have to recount his love of drawing, and even of writing verses. His own room was furnished with pictures, drawn, coloured, and neatly framed by himself, sometimes from copies, but often from life. Mr. Clark informed Dr. Stukely that the walls of the room in which St. Isaac lodged were covered with charcoal drawings of birds, beasts, men, ships, and mathematical figures, all of which were very well designed.

Among the portraits in Newton's room were those of Dr. Donne, Mr. Stokes, the master of Grantham school, and King Charles I, under whose picture were the following lines:

A secret art my soul requires to try,
If prayers can give me what the wars deny.
Three crowns distinguished here, in order do
Present their objects to my knowing view.
Earth's crown, thus at my feet I can disdain,
Which heavy is, and at the best but vain.
But now a crown of thorns I gladly greet,
Sharp is this crown, but not so sharp as sweet;
The crown of glory that I yonder see
Is full of bliss and of eternity.

These verses were repeated to Dr. Stukely by Mrs. Vincent, who believed them to be the composition of Newton, a circumstance which seems to be the more probable, as he himself many years after assured Mr. Conduit, with some expression of pleasure, that he "at one time excelled in making verses," though he had been frequently heard to express his great contempt for poetical composition; but like all young men who are anxious in gaining a respectable livelihood, he gave up the idle trade for a more serious calling.

But during these early years, while the mind of Newton was occupied principally with the pursuits which we have already detailed, it was by no means inattentive to the motions of the celestial bodies, on which he was afterwards enabled to throw such a flood of brilliant light. His thoughts had probably been directed to the more accurate measure of time which the motion of the sun afforded, by the imperfections of his water-clock. In the back-yard of Mr. Clark's house where he lived, he marked the varying movements of that luminary upon the walls and roofs of the outbuildings, and by means of fixed pins he noted the hourly and half-hourly subdivisions. One of these dials, which went by the name of "Isaac's dial," and was long after frequently referred to by the country people for the hour of the day, appears to have been drawn solely from the observations of several years; but we have now no means of obtaining information whether the numerous dials which he drew on the walls of his house at Woolsthorpe, and which existed many years after his death, were of the same description, or were projected from his knowledge of the doctrine of the sphere.

IN THE YEAR 1656, upon the death of the Reverend Mr. Smith, his widow, the mother of Newton, left the rectory of North Witham and took up her residence at Woolsthorpe along with her three children, Mary, Benjamin, and Hannah Smith. Isaac had now attained the fifteenth year of his age, and had made great progress in his studies under Mr. Stokes; and as his mother considered that he was capable of being useful in the management of the farm and country business at Woolsthorpe, she, chiefly from a motive of economy, recalled him from the school at Grantham, and brought him home. In order to accustom the young farmer to two of the most important branches of agricultural labour—the art of selling and buying—he was generally sent on the Saturdays to Grantham market, to dispose of grain and other farm produce, and to purchase such articles as were requisite for the use of the family. But as Isaac was yet void of experience in such matters, an old trust-worthy servant generally accompanied him for the purpose of initiating him into the mysteries of the business of market-making. The inn where they put up was the Saracen's Head at West Gate; but no sooner had they put up their horses than our youthful philosopher deserted his commercial concerns, and betook himself to his former lodgings in Mr. Clark's garret, where a number of the apothecary's old books afforded him an abundance of entertainment till his aged guardian had finished the more intricate business of market-making, and called for him with the announcement that it was time for them to return home. At other times he did not trouble himself by going so far, but deserted his duties on the road, and ensconced himself under a hedge, where he remained with some favourite volume till the servant returned from market at Grantham. "One of his uncles," says Mr. Biot, "having one day found him under a hedge with a book in his hand and entirely absorbed in meditation, took it from him, and found that he was occupied in the solution of a mathematical problem. Struck with finding so serious and so active a disposition at so early an age, he urged his mother no longer to thwart him, and to send him back to Grantham to continue his studies." This advice his mother shortly after found it necessary to comply with, as the more immediate affairs of the farm were not more prosperous under his management, than would have been his marketings at Grantham if they had been left to his own discretion. The perusal of a book, the execution of a model, or the superintendence of a water-wheel of his own construction, whirling the glittering spray from some neighbouring stream, absorbed all Isaac's thoughts, while the sheep were left to wander at will, and the cattle making sad havoc in the cornfields.

MRS. SMITH WAS SOON CONVINCED from experience that her son's genius took a higher flight than the cultivation of the soil, and as his passion for study, and his dislike for rural occupations increased with his years, she wisely resolved to give him all the advantage which education could confer. He was accordingly sent back to Grantham school, and his lodgings in Mr. Clark's house, where he continued for some months busily engaged in preparation for his academical studies. His uncle, the Reverend William Ayscough, who was rector of Burton Coggles, about three miles east of Woolsthorpe, and who had himself studied at Trinity College, recommended it to his nephew to enter that society, and it was accordingly determined that he should proceed to Cambridge at the commencement of the approaching term.

It has been justly remarked by a celebrated Scientific author that, "to a young mind thirsting for knowledge, and ambitious of the distinction which it brings, the transition from a village school to an university like that of Cambridge—from the absolute solitude of thought to the society of men imbued with all the literature and Science of the age, must be one of eventful interest. To Newton, it was a source of peculiar excitement. The history of Science affords many examples where the young aspirant had been early initiated into the mysteries, and had even exercised his powers of invention and discovery before he was admitted within the walls of a college; but he who was to give philosophy her laws did not exhibit such early talent; no friendly counsel regulated his youthful studies, and no work of Scientific eminence seems to have guided him in his course. In yielding to the impulse of his mechanical genius, his mind obeyed the laws of its own natural expansion, and, following the line of least resistance, it was thus drawn aside from the strongholds with which it was destined to grapple. When Newton, therefore, arrived at Trinity College, he brought with him a more slender portion of Science than falls to the lot of ordinary scholars; but this state of his acquirements was, perhaps, not unfavourable to the development of his powers. Unexhausted by premature growth, and invigorated by healthful repose, his mind was the better fitted to make those vigorous and rapid shoots which soon covered with foliage and with fruit, the genial soil to which it had been transferred. Cambridge was consequently the real birth-place of his (Newton's) genius. Her teachers fostered his earliest studies; her institutions sustained his mightiest efforts; and within her precincts were all his discoveries made and perfected. When he was called to higher official functions, his disciples kept up the pre-eminence of their master's philosophy, and their successors have maintained this seat of learning in the fullness of its glory, and rendered it the most distinguished among the universities of Europe."

~~JOHN THE BAPTIST PREACHING IN THE WILDERNESS~~

ISAAC BARROW TEACHING
ON THE TRINITY COLLEGE LANDHOLDINGS

Isaac Barrow did lecture throughout the extensive holdings of Trinity College, and preach the sacrament of education for the remission of ignorance.

ISAAC NEWTON WAS ADMITTED into Trinity College, Cambridge, on the 5th day of June, 1660, in the 18th year of his age. Dr. Barrow was elected professor of Greek in that university during the same year. Newton's attention was first directed to the study of mathematics by an ardent desire to inquire into the truth of judicial astrology, and he is said to have discovered the folly of that study by erecting a figure with the aid of one or two of the problems of Euclid. The propositions contained in this ancient system of geometry he regarded as self-evident truths; and, by his genius and patient application, without any preliminary study, he made himself master of Descartes' Geometry. He afterwards discovered, as a mistake in his mathematical studies, this neglect of the elementary truths of geometry; and he expressed to Dr. Pemberton his regret that "he had applied himself to the works of Descartes and other algebraic writers before he had considered the elements of Euclid with that attention which so excellent a writer deserved." Dr. Wallis's *Arithmetic of Infinities,* Saunderson's *Logic*, and the *Optics* of Kepler, were among the works which he appears to have studied with much care. On these books he wrote numerous comments during their perusal; and so great was his progress in his Scientific studies, that he is reported to have found himself more deeply versed in several branches of knowledge than the tutor who was employed to direct his studies.

During the first three years that he spent at Trinity College, neither history nor tradition has handed down to us any particular account of his progress. It appears from an account (which he regularly kept) of his expenses, that in the year 1664 he purchased a prism, for the purpose, as has been reported, of examining Descartes' Theory of Colours; and it is stated by Mr. Conduit, that he not only soon established his own views on the subject, but detected the errors in those of the French philosopher. This, however, does not appear to have been the case. Had he really made such an important discovery as the composition of light so early as the year 1664 or 1665, it is not at all probable that he would have withheld it not only from the Royal Society, but from his personal friends at Cambridge, till 1671. Dr. Barrow, his friend and tutor, was made Lucasian Professor of Mathematics in 1663, and he delivered a course of optical lectures, which he afterwards published in 1669. In the preface of this work he acknowledges his obligations to his esteemed colleague, Mr. Isaac Newton, for having revised the manuscripts and corrected several oversights, and made some important suggestions. In the twelfth lecture there are some observations on the origin and nature of colours, which Newton, had he been then in possession of their true theory, could not have permitted his friend to publish. According to Dr. Barrow, WHITE is that which discharges a copious light, equally clear in every direction. BLACK is that which does not emit light at all, or which does it very sparingly. RED is that which emits a light more clear than usual, but interrupted by shady interstices. BLUE is that which discharges a rarified light, as in bodies which consist of white and black particles arranged alternately. GREEN is nearly allied to blue. YELLOW is a mixture of much white and a little red; and PURPLE consists of a great deal of blue, mixed with a small portion of red. The blue colour of the sea arises from the whiteness of the salt which it contains, mixed with the blackness of the pure water in which the salt is dissolved; and the blueness of the shadows of bodies, seen at the same time by candle and daylight, arises from the whiteness of the paper mixed with the faint light or blackness of the twilight. These opinions savour so little of genuine philosophy that they must have attracted the observation of Newton; and had he discovered at that time, when he revised these manuscripts, that white was a mixture of all the colours, and black a privation of them all, he could not have permitted the absurd speculations of his master to pass uncorrected.

That Newton had not distinguished himself by any positive discovery, so early as 1664 or 1665, we think may be satisfactorily proved; but it may also be inferred from the circumstances which attended the competition for the law-fellowship of Trinity College. The only candidates for this appointment were Newton and Mr. Robert Uvedale; and, on examination, Dr. Barrow, then Master of

Trinity, having found them perfectly equal in their attainments, conferred the fellowship on Uvedale as the senior candidate. Had Isaac Newton made the important discovery alluded to, at this time, there can be little doubt that he would have received the appointment from his friend.

In the books of the University, we find Isaac Newton recorded as having been admitted sub-sizor in 1661. He became a scholar in 1664. In 1665, he took his degree of Bachelor of Arts; and in 1666, in consequence of the breaking out of the plague, he returned home to Woolsthorpe. In 1667, he was made Junior Fellow. In 1668, he took his degree of Master of Arts; and in the same year he was appointed to a Senior Fellowship. In 1669, when Dr. Barrow had resolved to devote his attention to theology, he resigned the Lucasian Professorship of Mathematics in favour of Newton, who may now be considered as having entered upon that brilliant career of discovery, which has, by general consent, placed his name at the head of those great men who have been the ornaments of their species. However imposing be the attributes with which time has invested the sages and the heroes of antiquity, the brightness of their fame has been eclipsed by the splendour of St. Isaac Newton's reputation; and neither the partiality of rival nations, nor the vanity of a presumptuous age, has ventured to dispute the ascendancy of his genius.

The life and writings of this great philosopher abound with the richest counsel. Here the philosopher will learn the art by which alone he can acquire an immortal fame. The moralist will trace the lineaments of a character adjusted to all the symmetry of which an imperfect nature is susceptible; and the religious man will contemplate with delight the High Priest of Science quitting the study of the material universe—the scene of his intellectual triumphs—to investigate, with humility and patience, the mysteries of his faith.

ABOUT THE AUTHOR

Mag. Augustina Comte is the High Priestess of the Temple of Science, a global fellowship of believers in Science, Technology, and the Spirit of Progress. Ordained in 2013, she holds an honorary doctorate in speculative theology from the Sapienza University of Rome.

In addition to her work on the Scientistic Scriptures series, Mag. Comte also travels widely for the purposes of research and evangelism. She has been invited to speak at such venues as the Banff Centre for Arts and Creativity (Alberta, Canada, 2015) and the Spencer Museum of Art at the University of Kansas (Lawrence, Kansas, U.S.A., 2018).

Outside of her professional roles, Mag. Comte enjoys collecting obscure facts and criticising people's grammar on the Internet. She resides in Utopia, Kansas, with her desktop, laptop, tablet, and smartphone. Follow the Magister on Facebook or Twitter to keep up to date with her latest publishing endeavours.

facebook.com/tosmagister

twitter.com/tosmagister

ABOUT THE PUBLISHER

In its ministry as the publishing arm of the Temple of Science, MAGISTERIUM PRESS produces rigorous scientific, philosophical, historical, and spiritual materials to support faith communities in spreading the good news of the Spirit of Progress. Mag. Augustina Comte, high priestess of the Temple of Science, serves as the senior editor at the press.

To meet the needs of the evolving Scientistic faith, MAGISTERIUM PRESS continually invests in the development of new resources by conducting research into the history and philosophy of Science and Technology in modern Western culture. Currently the major research focus is compiling and editing the *Book of Science*—the sacred text of Scientism. The Scientistic Scriptures series provides a format for the individual books that will ultimately make up this sacred anthology to be published as they are completed.

For more information about MAGISTERIUM PRESS, please visit magisterium.templeofscience.info.

ALSO BY MAG. AUGUSTINA COMTE

The Gospel of Isaac Newton

The Gospel of René Descartes (available Aug. 12)

Made in United States
North Haven, CT
10 November 2026

22080636R00143